Contents

A Note to Student	iv

Unit One
Chapter 1: Numbers for Algebra
Understanding Absolute Value	Exercise 1
Adding and Subtracting Integers	Exercise 2
Operations on Integers	Exercise 3
Using a Broken-Line Graph	Exercise 4
Working with Integers	Exercise 5
Solving Problems	Exercise 6

Chapter 2: Tools for Algebra
Using Properties of Addition and Multiplication	Exercise 7
Simplifying Expressions	Exercise 8
Finding the Value	Exercise 9
Properties	Exercise 10
Like Terms and Formulas	Exercise 11
Writing Equations	Exercise 12

Chapter 3: Solving Equations
Solving Equations	Exercise 13
Solving Equations	Exercise 14
Writing Equations	Exercise 15
Solving Equations	Exercise 16
Solving Equations	Exercise 17
Solving Equations and Problems	Exercise 18

Unit Two
Chapter 4: Introducing Functions
Coordinate Plane	Exercise 19
Functions	Exercise 20
Ordered Pairs	Exercise 21
Problem Solving	Exercise 22
Using a Bar Graph	Exercise 23

Chapter 5: Linear Equations and Functions
Using Slopes	Exercise 24
Graphing Equations	Exercise 25
Graphing Equations	Exercise 26
Using Slope	Exercise 27
Finding Intercepts	Exercise 28
Using Slope	Exercise 29
Using Rate of Change	Exercise 30

Chapter 6: Writing Linear Equations
Using Slope and y-Intercept	Exercise 31
Equations of Lines	Exercise 32
Equations of Lines	Exercise 33

Par	
Fin	
Writing Formulas	Exercise 36

Unit Three
Chapter 7: Inequalities
Graphing Inequalities	Exercise 37
Solving Inequalities	Exercise 38
Inequalities with Two Variables	Exercise 39
Graphing Solutions in the Coordinate Plane	Exercise 40
Using Inequalities	Exercise 41
Using Inequalities	Exercise 42

Chapter 8: Systems of Equations and Inequalities
Systems of Inequalities and Equations	Exercise 43
Solving Systems of Equations	Exercise 44
Solving Systems by Graphing	Exercise 45
Using Systems	Exercise 46
Solving Systems	Exercise 47
Using Systems	Exercise 48

Chapter 9: More About Data and Data Analysis
Representing Data	Exercise 49
Mean, Mode, Median, Range	Exercise 50
Displaying Data	Exercise 51
Describing Data	Exercise 52
Representing Data	Exercise 53
Quartiles	Exercise 54

Unit Four
Chapter 10: Exponents and Functions
Powers	Exercise 55
Finding Powers	Exercise 56
Division and Rules of Exponents	Exercise 57
Exponential Functions	Exercise 58
Using Tree Diagrams	Exercise 59
Using Exponents	Exercise 60

Chapter 11: Quadratic Functions and Equations
Graphing Quadratic Functions	Exercise 61
Minimums and Maximums	Exercise 62
Squares and Square Roots	Exercise 63
Quadratic Formula	Exercise 64
Writing Quadratic Equations	Exercise 65
Vertical Motion Formula	Exercise 66

Chapter 12: Polynomials and Factoring

Naming Polynomials	Exercise 67
Polynomials	Exercise 68
Factors	Exercise 69
Factoring	Exercise 70
Zero Products	Exercise 71
Using Formulas	Exercise 72

Unit Five
Chapter 13: Radicals and Geometry

Simplifying Radicals	Exercise 73
Operations with Radicals	Exercise 74
Solving Equations	Exercise 75
Right Triangles	Exercise 76
Right Triangles	Exercise 77
Using the Pythagorean Theorem	Exercise 78
Finding Distances	Exercise 79

Chapter 14: Rational Expressions and Equations

Rational Numbers and Expressions	Exercise 80
Simplifying Rational Expressions	Exercise 81
Rational Expressions	Exercise 82
Proportions	Exercise 83
Using Properties	Exercise 84
Inverse Variation	Exercise 85

Chapter 15: Topics from Probability

Probabilities	Exercise 86
Permutations and Combinations	Exercise 87
Probabilities	Exercise 88
Probability	Exercise 89
Events	Exercise 90
Probabilities	Exercise 91
Predicting Outcomes	Exercise 92

A Note to the Student

The exercises in this workbook go along with your *Pacemaker® Algebra 1* textbook. This workbook gives you the opportunity to review concepts, practice skills, and think critically.

Set goals for yourself and try to meet them as you complete each activity. The more you practice, the more you will remember. Being able to remember and apply information is an important skill, and leads to success on tests, in school, at work, and in life.

Your critical thinking skills will be challenged. You will need to think beyond what you learned in your textbook. The critical thinking activities provide you with the opportunity to put the information you have learned to use.

Your textbook is a wonderful source of knowledge. By completing the activities in this workbook, you will learn a great deal about algebra skills. The real value of the information will come when you have mastered these skills and put critical thinking to use.

PACEMAKER®

Algebra 1

Second Edition

WORKBOOK

GLOBE FEARON
Pearson Learning Group

Pacemaker® Algebra 1, Second Edition

We thank the following educators, who provided valuable comments and suggestions during the development of the First Edition of this book:

REVIEWERS

Ann Dixon, Teacher, Allegheny Intermediate Unit, South Park High School, Library, PA 15129

Ravi Kamat, Math Teacher, Dallas, TX 75203

Janet Thompson, Teacher, St. Louis, MO 63107

Linda White, Teacher, Riverview Learning Center, Daytona Beach, CA 32118

PROJECT STAFF

Executive Editor: Eleanor Ripp
Supervising Editor: Stephanie Petron Cahill
Senior Editor: Phyllis Dunsay
Editor: Theresa McCarthy
Production Editor: Travis Bailey
Lead Designer: Susan Brorein
Market Manager: Douglas Falk
Cover Design: Susan Brorein, Jennifer Visco
Editorial, Design, and Production Services: The GTS Companies
Electronic Composition: Phyllis Rosinsky

About the Cover: Algebra 1 is important in mathematics and in everyday life. The images on the cover represent some of the things you will be learning about in this book. The pattern in the nautilus shell is an example of a number pattern called the Fibonacci numbers. The graph of the equation $x = y$ shows how you can connect Algebra and Geometry. The running cheetah is an example of the familiar relationship between distance, rate, and time. The coins represent problems about discount, sale price, and the cost of items. The ruler is one of the tools you will use to measure the world around you. What other images can you think of to represent Algebra 1?

Copyright © 2001 by Pearson Education, Inc., publishing as Globe Fearon, an imprint of Pearson Learning Group, 299 Jefferson Road, Parsippany, NJ 07054. All rights reserved. No part of this book may be reproduced or transmitted in any form or by any means, electronic or mechanical, including photocopying, recording, or by any information storage and retrieval system, without permission in writing from the publisher. For information regarding permission(s), write to Rights and Permissions Department.

ISBN 0-130-23641-1
Printed in the United States of America
8 9 10   05

1-800-321-3106
www.pearsonlearning.com

Name _____ Date _____

 **Understanding Absolute Value**   **Exercise 1**

Lessons 1.1 and 1.2

Write *true* or *false* after each sentence. If the sentence is *false*, change the underlined word or words to make it true.

1. Numbers to the left of 0 on the number line are <u>positive</u>.

   _____

2. The distance of a number from 0 is <u>always</u> positive.

   _____

3. Five is the <u>absolute value</u> of both 5 and ⁻5.

   _____

4. ⁻7 is <u>greater</u> than 4.

   _____

5. 0 is <u>less</u> than ⁻3.

   _____

CRITICAL THINKING

Find the integers.

1. List the integers with absolute value less than 2. _____

2. List the negative integers with absolute value less than 5. _____

3. List the positive integers with absolute value less than 4. _____

4. List the integers with absolute value 7. _____

Chapter 1 • Numbers for Algebra   1

Name _____ Date _____

| **1** | **Adding and Subtracting Integers** | **Exercise 2** |

Lessons 1.1, 1.3, and 1.4

**A. Tell which direction from zero you would move to graph
the integer. Write *left* or *right*.**

   1. $^-2$ _____

   2. 3 _____

   3. $^+2$ _____

   4. 0 _____

   5. $^-(8)$ _____

   6. $|^-4|$ _____

B. Add using the number line.

   1. $3 + 3$

   2. $^-5 + 2$

   3. $4 + (^-2)$

   4. $^-3 + 0$

C. Add or subtract.

   1. $7 + (^-2)$

   2. $^-3 - 8$

   3. $^-5 + 1$

   4. $^-(4) - 6$

   5. $0 - (^-9)$

   6. $^-2 - (^-5)$

   7. $^-8 + |6|$

   8. $|^-4| + |^-3|$

   9. $5 - |^-4|$

   10. $|6| + (^-1)$

   11. $(^-1) - |^-3|$

   12. $(^-2) + |0|$

Chapter 1 • Numbers for Algebra

Name _____   Date _____

 **Operations on Integers**   **Exercise 3**

Lessons 1.5, 1.6, and 1.8

A. Write division facts to complete the table.

1. $^-2 \bullet 5 = {}^-10$	$^-10 \div (^-2) = 5$	$^-10 \div 5 = {}^-2$
2. $6 \bullet (^-3) = {}^-18$		
3. $(^-1)(^-9) = 9$		
4. $4 \bullet (^-2) = {}^-8$		

B. Multiply or divide.

1. $^-4 \bullet 6$

2. $(^-20) \div (^-5)$

3. $^-3 \bullet 8$

4. $\dfrac{^-5}{5}$

5. $(7)(^-2)$

6. $\dfrac{^-14}{^-2}$

7. $0 \div 9$

8. $^-2 \bullet 5$

9. $12 \div (^-4)$

10. $(^-2) \bullet (^-2)$

11. $6 \bullet (^-3)$

12. $\dfrac{6}{^-1}$

C. Find the power.

1. $(^-2)^2$

2. $(^-2)^3$

3. $(^-2)^4$

4. $^-7^2$

5. $(^-1)^8$

6. $(^-3)^3$

Chapter 1 • Numbers for Algebra   3

Name _____ Date _____

 **1** ▷ **Using a Broken-Line Graph**   **Exercise 4**

Lesson 1.11

Use the graph to answer the questions.

A hot air balloon at 500 feet drops at a constant rate of 25 feet per second for 10 seconds. It stops for 5 seconds, rises 50 feet at a constant rate of 10 feet per second, and then drops to the ground at a constant rate of 20 feet per second.

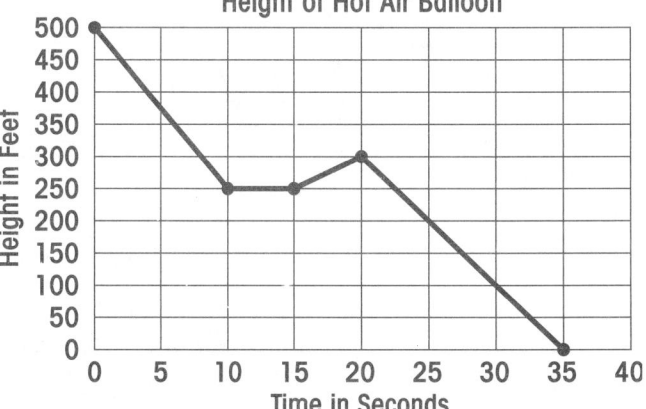

1. After the first 10 seconds, how far has the balloon dropped?

2. How high is the balloon when it stops for 5 seconds?

3. How many seconds total does it take the balloon to drop to a height of 200 feet?

4. How many seconds does it take for the balloon to land?

5. How many seconds does it take for the balloon to fall to the ground?

4   Chapter 1 • Numbers for Algebra

Name _____ Date _____

 **1** ▶ **Working with Integers** Exercise 5

Lessons 1.2 and 1.7

A. Complete the table. Give the opposite in words. Then, write the opposite as an integer.

	Opposite in words	Integer
1. Earned $12	Spent $12	⁻12
2. Spent $50		
3. 20°C rise in temperature		
4. Down 5 feet		
5. Up 28 feet		

B. Simplify.

1. $3 - {}^-4 + 10$ **2.** ${}^-6 + 3 - 10$ **3.** ${}^-3 - 14 - 7$

4. $3 + {}^-4 - 10$ **5.** ${}^-5 - {}^-5 - 5$ **6.** ${}^-6 - 7 - {}^-13$

7. $14 + 7 - {}^-3$ **8.** ${}^-10 - {}^-4 - 2$ **9.** $6 - 2 - 8$

CRITICAL THINKING

Solve the problem. Show your work.

Reina heard on the 6:00 P.M. news that the temperature had dropped 22° since 4:00 P.M. At 4:00 P.M., the temperature was 12°. What is the temperature at 6:00 P.M.?

Chapter 1 • Numbers for Algebra   5

Name _____ Date _____

 1 ▶ **Solving Problems**   **Exercise 6**

Lesson 1.10

Solve each problem using Guess, Check, Revise.

1. In a fishing contest, the two largest bass weighed a total of 22 pounds. The first-place bass weighed 8 pounds more than the second-place bass. Find the weights of the first-place and second-place bass.

2. The band members sold 316 juice drinks for a fund-raiser. They sold 52 more grape drinks than strawberry drinks. How many of each type of juice drink did they sell?

3. The product of two integers is 50. One integer is twice the other. Find the integers.

4. The sum of two integers is ⁻1. The product of the integers is ⁻72. Find the integers.

6   Chapter 1 • Numbers for Algebra

Name _____   Date _____

 **2** **Using Properties of Addition and Multiplication**   **Exercise 7**

Lessons 2.1 to 2.4, 2.8, and 2.10

A. Write *true* or *false* after each sentence. If the sentence is *false,* change the underlined word or words to make it true.

   1. In the expression $7x + 15$, 15 is a <u>coefficient</u>.

       _____

   2. $\dfrac{3x + 7}{2}$ means <u>$(3x + 7) \div 2$</u>.

       _____

   3. You can rewrite $2(4 + 8)$ as $(2)(4) + (2)(8)$ using the <u>Distributive Property</u>.

       _____

   4. An example of the <u>Associative Property</u> is $(3 + 7) + 14 = 3 + (7 + 14)$.

       _____

   5. To show 9 <u>decreased by</u> 6, you can write $9 - 6$.

       _____

B. Find the value of each number expression.

   1. $12 - (2 + 6)$ _____   **2.** $2 + 3 \times 8$ _____

       _____   _____

   3. $5(4 + 8)$ _____   **4.** $14 + 3^3 \times 3$ _____

       _____   _____

Chapter 2 • Tools for Algebra   7

Name _____ Date _____

2 ▶ Simplifying Expressions

Lessons 2.1, 2.5, 2.7, 2.10, and 2.11

Exercise 8

A. Simplify each expression. Circle the correct answer.

1. $3(x + 7)$

    a. $3x + 7x$ **b.** $3x + 21$

    c. $7x + 21$ **d.** cannot be simplified

2. $5x + 6 + 2y + 3x + 4y$

    a. $14xy + 6$ **b.** $6 + 7xy + 7xy$

    c. $8x + 6y + 6$ **d.** cannot be simplified

3. $2x + (3 + 5x)$

    a. $2x + 8$ **b.** $7x + 3$

    c. $10x$ **d.** cannot be simplified

4. $2x + 3y + 7 + z$

    a. $7 + 6xyz$ **b.** $12xy + z$

    c. $(2x + 3y + z) + 7$ **d.** cannot be simplified

5. $4 + 33 + 4 \times 2$

    a. 45 **b.** 82

    c. 78 **d.** 43

B. Are the two expressions equivalent? Write *yes* or *no*.
   Show your work.

    1. $2x + 6x$ and $x(2 + 6)$ _____   **2.** $3 + (4 \times 3)$ and $(3 + 4) \times 2$ _____

    3. x^5 and $x \cdot x \cdot x \cdot x \cdot x$ _____   **4.** $4 - 11$ and $11 - 4$ _____

8   Chapter 2 • Tools for Algebra

Name _____ Date _____

2 ▶ Finding the Value   Exercise 9

Lessons 2.3 and 2.6

Write the expressions. Then evaluate.

1. **a.** the product of 5 and a number x _____

   b. Evaluate when $x = -1$.

2. **a.** 18 decreased by a number z _____

   b. Evaluate when $z = 23$.

3. **a.** the quotient of 16 and a number m _____

   b. Evaluate when $m = -4$.

4. **a.** the product of 8 and twice a number n _____

   b. Evaluate when $n = 3$.

5. **a.** the sum of 3 times a number k and 4 _____

   b. Evaluate when $k = -2$.

Chapter 2 • Tools for Algebra   9

Name _____ Date _____

 2 ▶ **Properties** Exercise 10

Lessons 2.8 to 2.11

A. Which equation shows the property? Circle the correct answer.

1. Addition Property of Opposites

    a. $2 + 3 = 3 + 2$

    b. $2 + (3 + 4) = (2 + 3) + 4$

    c. $2 + -2 = 0$

2. Associative Property of Multiplication

    a. $3(2 \times 4) = 3(2) \times 3(4)$

    b. $3(2 \times 4) = (3 \times 2)4$

    c. $3 \times 5 = 5 \times 3$

3. Distributive Property

    a. $7(x + y) = 7x + 7y$

    b. $7(x + y) = (7x)(7y)$

    c. $7(x + y) = (x + y)7$

4. Identity Property of Multiplication

    a. $x \bullet 0 = 0$

    b. $x \bullet x = x^2$

    c. $x \bullet 1 = x$

B. Simplify.

1. $7(x + 9)$

2. $7z + z(2 + 5)$

3. $24x + (3x + 17)$

4. $3 + 7(x + 5)$

10   Chapter 2 • Tools for Algebra

Copyright © by Globe Fearon Inc. All rights reserved.

Name _____ Date _____

2 ▶ Like Terms and Formulas   Exercise 11

Lessons 2.4 and 2.14

A. Match like terms. Write the correct letters on the lines.

_____ **1.** $3x^2$ **a.** a^2b

_____ **2.** $2ab$ **b.** $10x$

_____ **3.** $-5x$ **c.** $2a$

_____ **4.** a **d.** $-3x^2$

_____ **5.** $-4a^2b$ **e.** $2ba$

B. Match each picture of the figure with its formula.
Write the letter on the line.

1. _____

5 in.   5 in.

8 in.

a. $A = lw$

$A = 5 \cdot 8$

2. _____

8 in.

5 in.   5 in.

8 in.

b. $V = lwh$

$V = 8 \cdot 5 \cdot 5$

3. _____

5 in.

5 in.

8 in.

c. $P = a + b + c$

$P = 5 + 5 + 8$

Copyright © by Globe Fearon Inc. All rights reserved.

Chapter 2 • Tools for Algebra   11

Name _____ Date _____

2 ▶ Writing Equations   Exercise 12

Lessons 2.13 and 2.14

Write an equation for each problem.

1. Tickets to a play cost $6.50 each. Write an equation for the total cost of 12 tickets plus a $7.50 fee for large groups.

2. A shirt order consists of 10 small, 5 medium, and 8 large shirts. The prices of the shirts are small $5.00; medium $7.50; large $12.00. There is a mail order charge of $.50 per shirt for shipping and handling. Write an equation for the total cost of ordering the shirts by mail.

3. The total cost of ribbon is the product of the total number of yards and the cost per yard. The cost per yard is $.40. Write an equation for the total cost of the following:

   2 yards blue ribbon

   8 yards white ribbon

   11 yards pink ribbon

   7 yards peach ribbon

CRITICAL THINKING

Solve the problem. Show your work.

Find the total area of the flower and vegetable garden. Use the formula for the area of a rectangle.

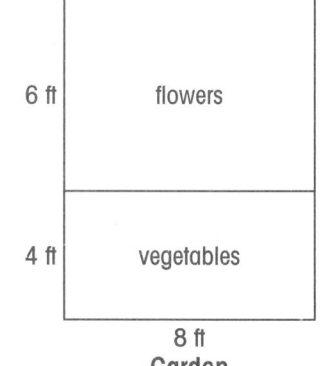

12   Chapter 2 • Tools for Algebra

Name _____ Date _____

 3 **Solving Equations** **Exercise 13**

Lessons 3.1 to 3.4, 3.6, 3.8, and 3.9

A. Is the first number a solution for the equation? Write *yes* or *no*. Show your work.

1. -5; $3z + 4 = 15$

2. 16; $8 - \dfrac{x}{2} = 0$

3. 12; $4(p + 5) = 60$

4. -40; $8 - k = -48$

B. Circle the correct answer for each question.

1. Which equation has the same solution as $4x = 16$?

   a. $x = 8$ b. $x = 4$

   c. $x = -32$ d. $x = 64$

2. Which operation is needed to solve $6a = 36$?

   a. addition   b. subtraction

   c. multiplication   d. division

3. Which operation is needed to solve $m - 8.83 = 12.75$?

   a. addition   b. subtraction

   c. multiplication   d. division

4. What is the first step you use to solve $15 = 6y - 9$?

   a. add 9   b. divide by 6

   c. subtract 15   d. multiply by 15

Chapter 3 • Solving Equations   13

Name _____ Date _____

3 ▶ Solving Equations

Lessons 3.1 and 3.4

Exercise 14

A. Evaluate the equation. Answer the question.

Renelle solved the equation $60 = 12n$ and found a solution of 720.

Did Renelle get the right solution? _____

Explain how you decided if Renelle's solution was correct. _____

_____

_____

_____

_____

B. For each equation, give the following information.

Name the operation in the equation.
Name the inverse of that operation.
Solve by showing a step that uses the inverse operation.

1. $d + 7 = 5$

    operation: _____

    inverse: _____

    solve:

2. $t - 57 = 39$

    operation: _____

    inverse: _____

    solve:

3. $6 = \dfrac{a}{2}$

    operation: _____

    inverse: _____

    solve:

4. $-3x = 12$

    operation: _____

    inverse: _____

    solve:

Name _____ Date _____

 **3** ▶ **Writing Equations**   **Exercise 15**

Lessons 3.5 to 3.8

Choose the equation that describes the problem.

1. Joaquin weighs his dog by weighing himself while
    holding the dog. Their total weight is 237 pounds.
    Joaquin weighs 156 pounds. Use d for the dog's weight.
    Which equation shows the dog's weight?

    a. $d + 237 = 156$ **b.** $d - 156 = 237$ **c.** $d + 156 = 237$

2. A baseball diamond has four equal sides. The total
    distance around the diamond is 360 feet. Use s for side.
    Which equation shows the length of a side?

    a. $\frac{s}{4} = 360$ **b.** $4s = 360$ **c.** $s + 4 = 360$

CRITICAL THINKING

Write an equation for each problem. Then solve.

1. The headquarters of the United States Department of
    Defense is the Pentagon. The Pentagon has 5 sides that
    are all the same length. If the perimeter of the Pentagon
    is 1,600 m, what is the length of each side?

2. The total weight of Maya and her large cat is 157 pounds.
    The cat weighs 19 pounds. How much does Maya weigh?

Chapter 3 • Solving Equations   15

Name _____ Date _____

 3 **Solving Equations**   **Exercise 16**

Lessons 3.1, 3.2, 3.4, 3.7, and 3.9 to 3.11

A. Write *true* or *false* after each sentence. If the sentence is *false,* change the underlined word or words to make it true.

1. Equivalent equations are equations that have the same <u>variable</u>.

_____

2. The <u>solution</u> to an equation is any value of the variable that makes the equation true.

_____

3. <u>Inverse operations</u> are operations that "undo" each other. For example, addition and subtraction are inverse operations.

_____

4. Discount is the amount you <u>spend</u> when you buy an item on sale.

_____

B. Solve each equation. Then, check the solution.

1. $5(12 + k) = 215$ **2.** $\frac{y}{7} = -7$

3. $36 = -9(w + 1)$ **4.** $42 + 12d = 5d$

16   Chapter 3 • Solving Equations

Name _____ Date _____

 3 ▷ **Solving Equations**   **Exercise 17**

Lessons 3.5 to 3.11

Solve each equation. Show all your steps. Check.

1. $12x - 18 = 126$ **2.** $-8(5 + k) = -104$

3. $36 + 3m = 12m$ **4.** $180 = \dfrac{v}{2} + 4$

5. $\dfrac{x}{-5} + 6 = -4$ **6.** $12(r - 6) = 48$

CRITICAL THINKING

Solve each problem. Show your work.

1. The football team scored 24 points in the first half. The team's final score was 45 points. How many points did the football team score in the second half?

2. Lonna has $278.00 in her savings account. Debbie has 3 times as much as Lonna. How much money does Debbie have in her savings account?

3. Harry earned $171.00 last week. He worked 30 hours. What is his hourly pay rate?

Copyright © by Globe Fearon Inc. All rights reserved.

Chapter 3 • Solving Equations   **17**

Name _____ Date _____

 3 ▶ **Solving Equations and Problems**   **Exercise 18**

Lessons 3.5, 3.6, 3.13, and 3.14

Write an equation for each problem. Then solve the problem.

1. 75% of the tickets to the concert in the auditorium were sold. The auditorium has 1,200 seats. How many tickets were sold?

2. Wei Lu needs new tires. Tires are regularly priced at $48.00 each, but they are on sale for 20% off the regular price. What is the sale price of tires?

3. Janine bought a new blouse for 25% off. The blouse was originally $28.00. What was the sale price?

CRITICAL THINKING

Solve each problem. Show your work.

1. Maria started with $58.00 in her checking account. She made a deposit of $347.00 and wrote a check for $265.00 to pay her rent. She wrote another check for groceries. Maria has $92.00 left in her account. What was the amount of the check for groceries?

2. Jamal earns $6.00 an hour for working 40 hours a week. If he works more than 40 hours in a week, he makes $9.00 an hour for every hour over 40. Last week, Jamal worked 48 hours. How much did he earn?

18   Chapter 3 • Solving Equations

Name _____   Date _____

4 ▶ Coordinate Plane   Exercise 19

Lessons 4.1 and 4.2

A. Write *true* or *false* after each sentence. If the sentence is *false,* change the underlined word or words to make it true.

1. The order of the numbers in an ordered pair <u>is not</u> important.

_____

2. The <u>horizontal</u> axis is called the *x*-axis.

_____

3. The point where the coordinate axes cross is called the <u>center</u>.

_____

B. Give the ordered pair of each point.

1. Point *A* _____

2. Point *B* _____

3. Point *C* _____

4. Point *D* _____

5. Point *E* _____

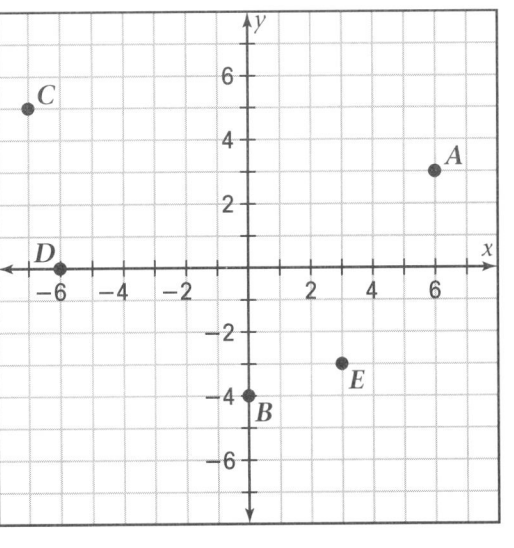

C. On the grid above, graph and label each point.

1. *F* at (2, 3)   **2.** *G* at (3, 2)   **3.** *H* at (−4, −3)   **4.** *I* at (6, −4)

Chapter 4 • Introducting Functions   19

Name _____ Date _____

 4 ▶ **Functions**   **Exercise 20**

Lessons 4.3 and 4.4

A. Graph and label the ordered pairs.

1.

Point	Time	Temperature
A	0	−1
B	1	−3
C	2	−5
D	3	−4
E	4	0

B. Complete each table.

1. $y = x + 1$

x	x + 1	y
−2		
−1		
0		
1		
2		

2. $y = -2x$

x	−2x	y
−2		
−1		
0		
1		
2		

3. $y = 3x + 2$

x	3x + 2	y
−2		
−1		
0		
1		
2		

4. $y = 2(x - 3)$

x	2(x − 3)	y
−2		
−1		
0		
1		
2		

20   Chapter 4 • Introducting Functions

Name _____ Date _____

  4 ▶ **Ordered Pairs**   **Exercise 21**

Lessons 4.5 and 4.6

A. Find $f(5)$ for the following functions.

  1. $f(x) = -x$ **2.** $f(x) = 3x^2$ **3.** $f(x) = 4(x + 1)$

  $f(5) =$ _____ $f(5) =$ _____ $f(5) =$ _____

B. Use the vertical line test to determine whether the graph is a function. Write *yes* or *no* on the line.

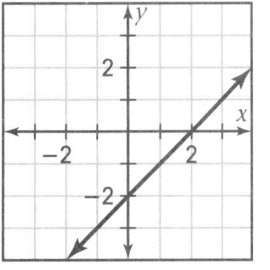   **1.** _____

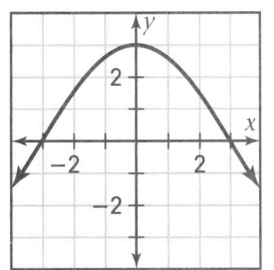   **2.** _____

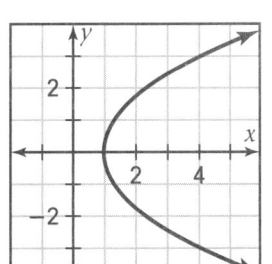   **3.** _____

Chapter 4 • Introducting Functions   21

Name _____ Date _____

 4 ▸ **Problem Solving**   **Exercise 22**

Lesson 4.8

A. Complete the table of ordered pairs.

x	$10x + 4$	y
1	$10(1) + 4$	
2		
4		
7		
10		

B. Use the above table to answer the questions.

You want to order posters by mail. Each poster costs $10.00, and you have to pay $4.00 for shipping. Let x be the number of posters you buy and y the total cost of the order.

   1. What is the equation that you would use to show the total cost?

   2. How much do 4 posters cost, including shipping?

   3. How much do 15 posters cost, including shipping?

   4. If you order 20 posters at once, does it cost the same, including shipping, as two separate orders of 10 posters each? Explain your answer.

22   Chapter 4 • Introducting Functions

Name _____ Date _____

 **4** ▶ **Using a Bar Graph**   **Exercise 23**

Lesson 4.9

Use the bar graph below to answer the questions.

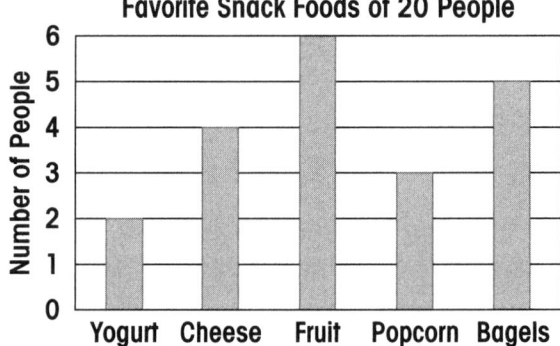

1. How many people reported that bagels were their favorite snack?

2. Which snack was favored by the most people?

3. Look at yogurt on the bar graph. Which snack was favored by twice as many people as yogurt?

4. Which snack was the second favorite?

5. How many people reported that cheese was their favorite snack?

6. How many more people reported popcorn than yogurt as their favorite snack?

Chapter 4 • Introducting Functions   23

Name _____ Date _____

5 ▶ Using Slopes

Exercise 24

Lessons 5.1 to 5.3, 5.6, and 5.9

A. Write *true* or *false* after each sentence. If the sentence
is *false,* change the underlined word or words to make
it true.

1. The point where a line crosses the *x*-axis is called the <u>*y*-intercept</u>.

_____

2. The <u>standard</u> form of the equation of a line is $y = mx + b$.

_____

3. <u>Rise</u> is the change between two points on a line in an up-and-down direction.

_____

4. A linear equation is an equation whose graph is a <u>circle</u>.

_____

B. Tell whether the ordered pair is a solution of $y = 4 - 2x$.
Write *yes* or *no.* Show your work.

1. $(0, -2)$ _____   **2.** $(2, 1)$ _____   **3.** $(1, 2)$ _____

4. $(-1, 5)$ _____   **5.** $(3, -2)$ _____   **6.** $(0, 4)$ _____

Name _____ Date _____

 5 ▶ **Graphing Equations**   **Exercise 25**

Lesson 5.2

Make a table of values to show ordered pairs for each
equation. Then, graph the ordered pairs.

1. $y = 3 - x$

x	$3 - x$	y
−2		
−1		
0		
1		
2		

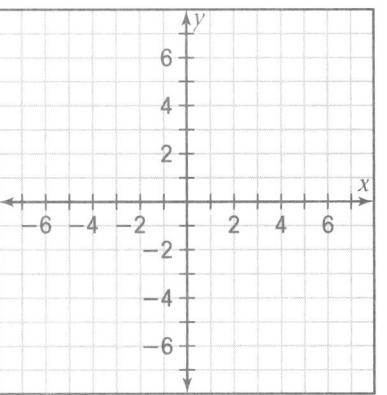

2. $y = x + 4$

x	$x + 4$	y
−2		
−1		
0		
1		
2		

3. $y = -4 + 2x$

x	$-4 + 2x$	y
−2		
−1		
0		
1		
2		

Chapter 5 • Linear Equations and Functions   25

Name _____ Date _____

5 ▶ Graphing Equations   Exercise 26

Lessons 5.2 to 5.4

A. Find at least three points, and graph each equation.

    1. $y = 3x - 3$

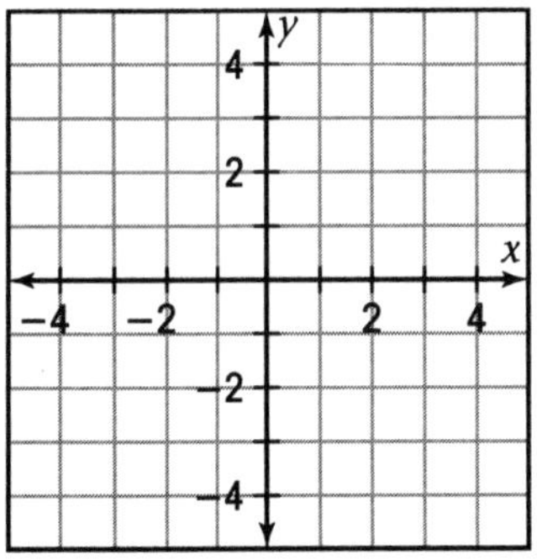

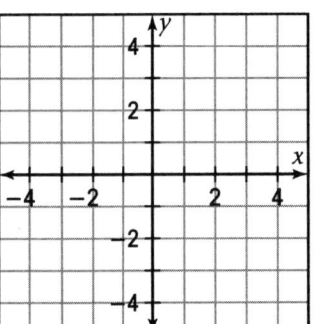

    2. $y = 2 - 2x$

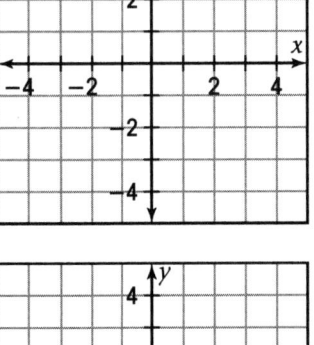

    3. $y = 3$

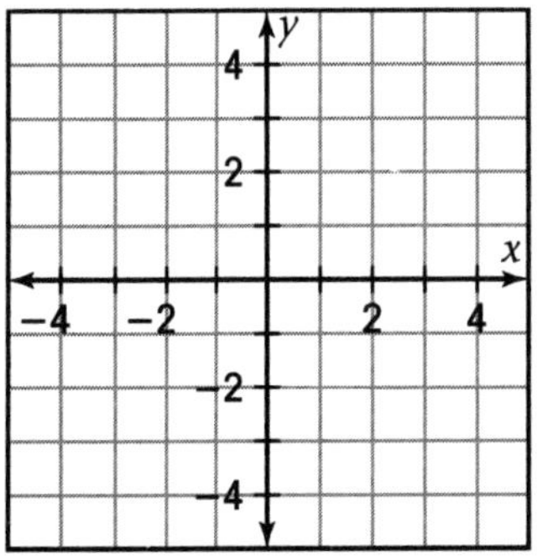

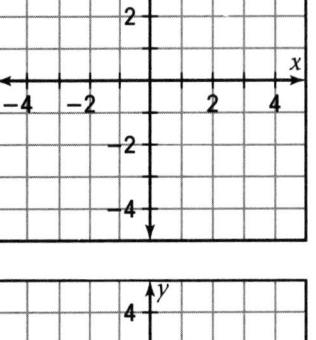

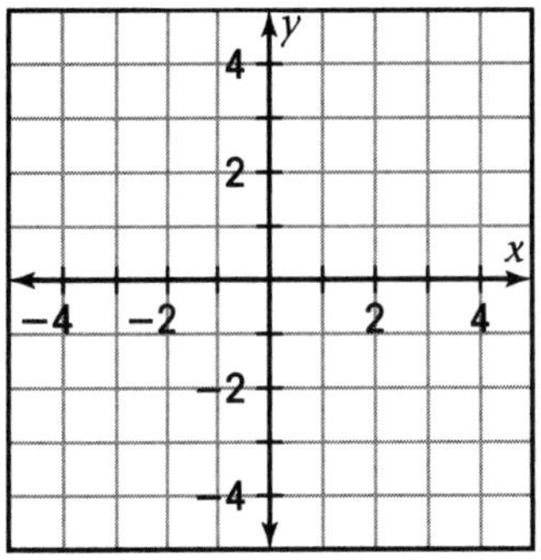

B. Find the slope of the line that contains the given points.

    1. (3, 1) and (4, 7)   **2.** (−2, 6) and (1, 3)

       slope _____   slope _____

    3. (3, 4) and (0, 2)   **4.** (5, −1) and (4, −1)

       slope _____   slope _____

26   Chapter 5 • Linear Equations and Functions

Name _____ Date _____

5 ▷ Using Slope   Exercise 27

Lesson 5.5

A. Tell whether the lines containing these pairs of points
are parallel.

  1. line 1: (6, 4) and (4, 2)   **2.** line 1: (0, 1) and (3, 10)

  line 2: (9, 7) and (6, 4)   line 2: (7, 3) and (2, 1)

  3. line 1: (8, 8) and (4, −4)   **4.** line 1: (1, −7) and (−4, 3)

  line 2: (3, −6) and (7, 2)   line 2: (−2, 8) and (1, 2)

B. Tell whether the lines containing these pairs of points
are perpendicular.

  1. line 1: (12, 3) and (8, 4)   **2.** line 1: (6, 7) and (4, 8)

  line 2: (1, 9) and (2, 5)   line 2: (6, 8) and (5, 6)

  3. line 1: (10, 8) and (5, −2)   **4.** line 1: (−4, −5) and (0, 3)

  line 2: (5, 3) and (4, 1)   line 2: (5, −3) and (−5, 2)

Chapter 5 • Linear Equations and Functions   27

Name _____ Date _____

5 ▷ Finding Intercepts

Exercise 28

Lesson 5.6

**A. Use the graph to find the *x*-intercept and *y*-intercept of
each line.**

1.

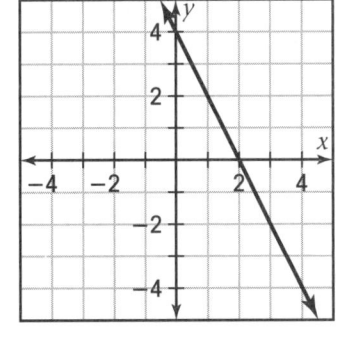

x-intercept _____

y-intercept _____

2.

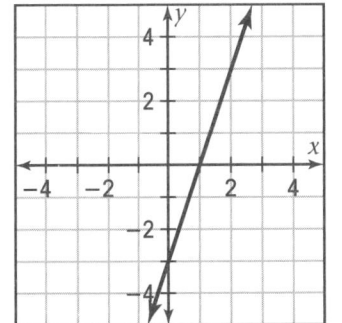

x-intercept _____

y-intercept _____

**B. Find the *x*-intercept by substituting 0 for *y*. Then find the
y-intercept by substituting 0 for *x*.**

1. $y = 6x - 6$

x-intercept _____

y-intercept _____

2. $y = 4 - 2x$

x-intercept _____

y-intercept _____

3. $y = x + 5$

x-intercept _____

y-intercept _____

4. $y = 7$

x-intercept _____

y-intercept _____

28   Chapter 5 • Linear Equations and Functions

Copyright © by Globe Fearon Inc. All rights reserved.

Name _____ Date _____

5 ▷ Using Slope

Exercise 29

Lessons 5.7 to 5.9

A. Graph the line that contains the given point and has the given slope.

1. point: $(-6, 2)$; slope: $\dfrac{2}{3}$

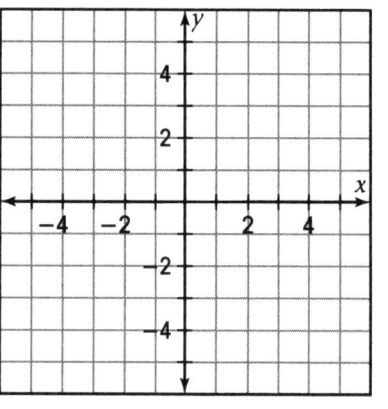

2. point: $(5, 0)$; slope: -1

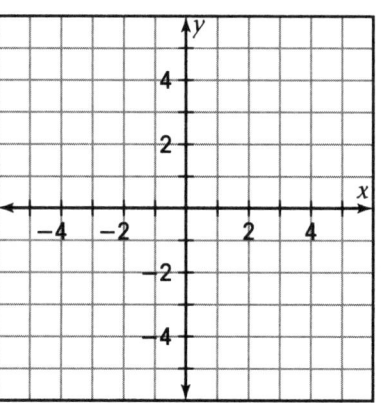

3. point: $(-4, 4)$; slope: $\dfrac{1}{4}$

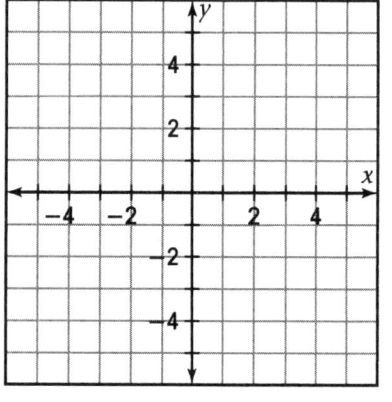

4. point: $(1, -2)$; slope: -3

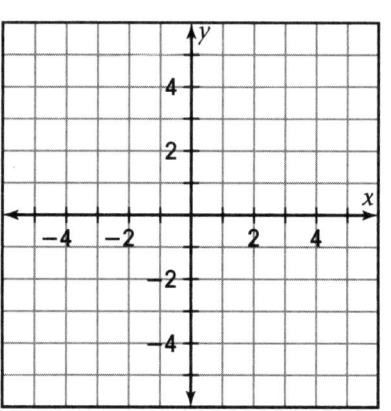

B. Write these equations in slope-intercept form. Then, find the slope and *y*-intercept of each line.

1. $6x + 4y = -12$

equation _____

slope _____

y-intercept _____

2. $-6x + 3y = 18$

equation _____

slope _____

y-intercept _____

Chapter 5 • Linear Equations and Functions   29

Name _____ Date _____

5 ▷ Using Rate of Change   Exercise 30

Lessons 5.11 and 5.12

A. Graph and find the slope to solve each problem.

1. Marlene is selling popcorn at a football game. At 6:00 P.M., she has sold 20 bags of popcorn. By 8:00 P.M., she has sold 60 bags. At this rate, how many bags of popcorn will she have sold by 10:00 P.M.?

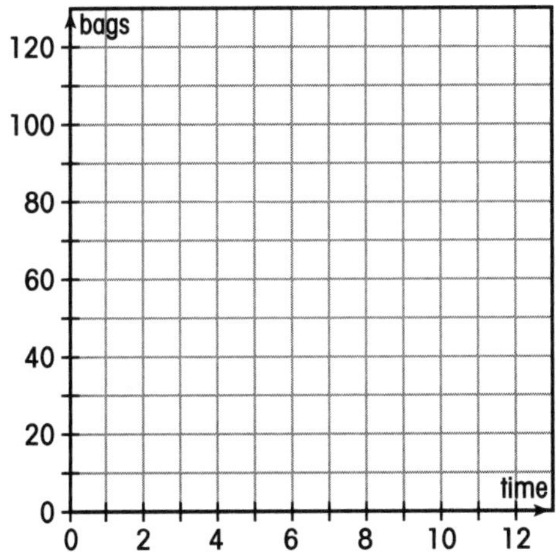

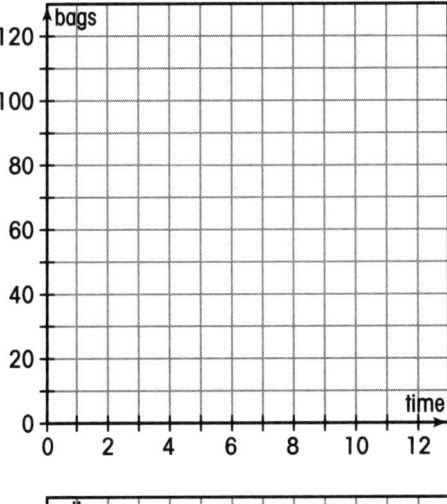

2. A train leaves the station at 5:00 A.M. By 8:00 A.M. it has traveled 300 miles. At this rate, how many miles will the train have traveled by 10:00 A.M.?

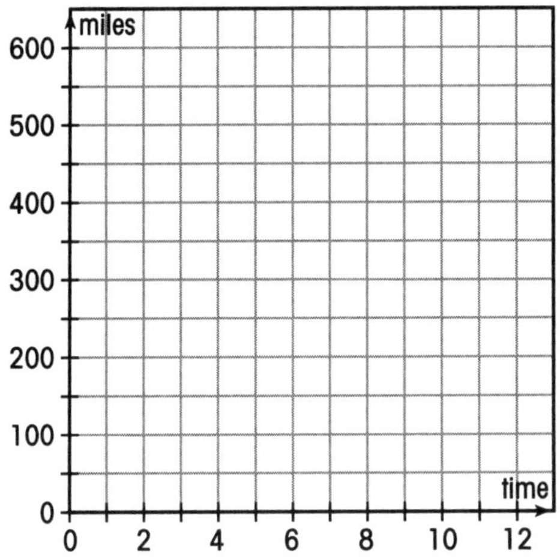

B. Robert is driving 240 kilometers to the beach. The distance Robert drives varies directly with the hours he drives. He has driven 120 kilometers in 3 hours.

1. Find the r in the equation $D = rt$ (Distance = rate × time)

2. Keeping that pace, how long will it take Robert to drive to the beach?

30   Chapter 5 • Linear Equations and Functions

Copyright © by Globe Fearon Inc. All rights reserved.

Name _____ Date _____

6 ▶ Using Slope and *y*-Intercept

Lesson 6.1

Exercise 31

A. Write the equation of the line with the given slope and *y*-intercept.

1. slope = 4 and *y*-intercept = −2 _____

2. slope = 0 and *y*-intercept = 10 _____

3. slope = −3 and *y*-intercept = 6 _____

4. slope = 5 and *y*-intercept = 0 _____

5. slope = $\frac{2}{3}$ and *y*-intercept = 9 _____

B. Use the graph to write the equation of each line.

1. _____

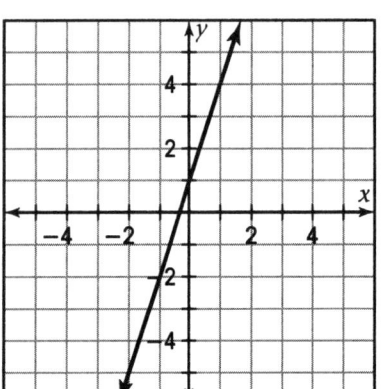

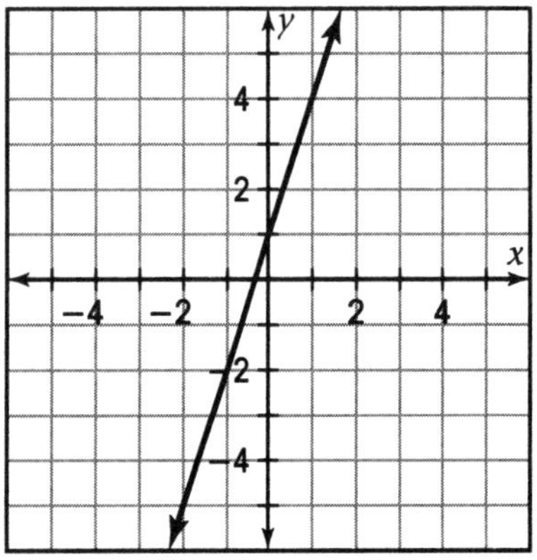

2. _____

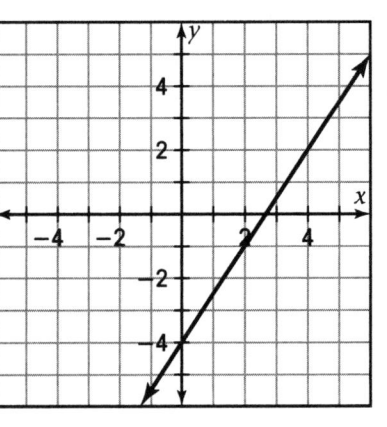

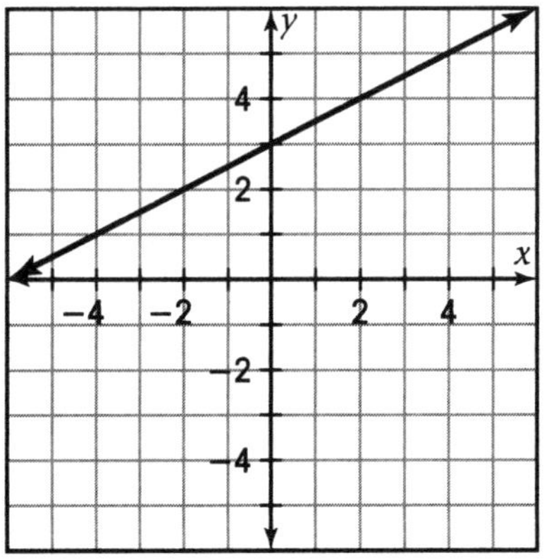

3. _____

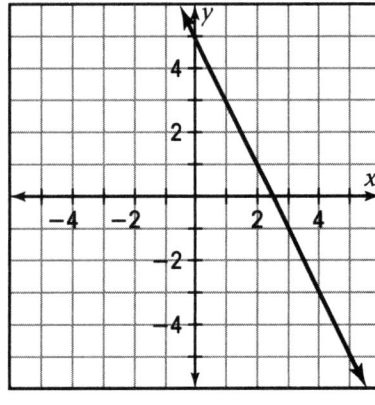

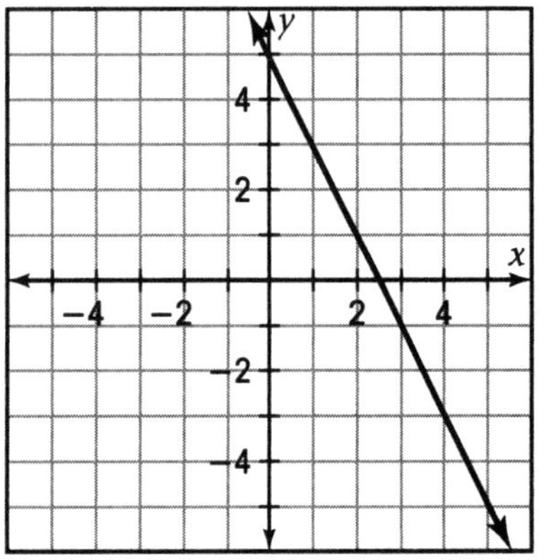

4. _____

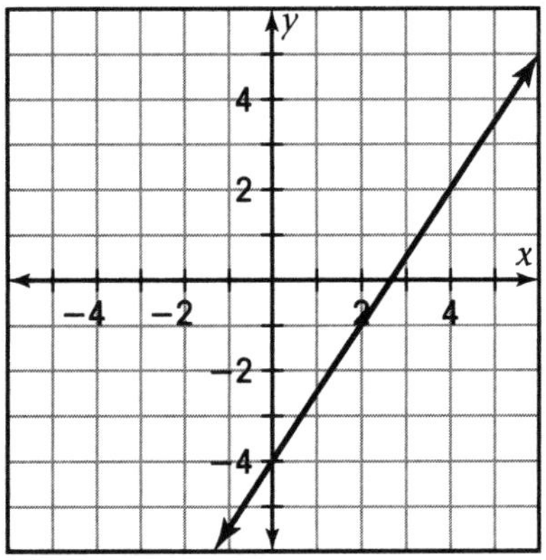

Copyright © by Globe Fearon Inc. All rights reserved.

Chapter 6 • Writing Linear Equations   31

Name _____ Date _____

6 ▶ Equations of Lines
Exercise 32

Lessons 6.2 and 6.6

A. Does the line pass through the given point? Write *yes* or *no* on the line. Show your work.

    1. $(4, 2)$; $y = 2x + 2$ _____

    2. $(1, -1)$; $y = -x$ _____

    3. $(5, 3)$; $y = x - 2$ _____

    4. $(0, 7)$; $y = x + 6$ _____

B. Write the equation of the line with the given point and slope. Then check the equation.

    1. point: $(4, 3)$ slope: 2 equation _____

    2. point: $(3, 2)$ slope: $\dfrac{2}{3}$ equation _____

    3. point: $(5, 1)$ slope: -3 equation _____

32   Chapter 6 • Writing Linear Equations

Name _____ Date _____

6 ▶ Equations of Lines

Lessons 6.1, 6.3, and 6.4

Exercise 33

A. Find the equation of the line through the given points.
Then check the equation.

1. (3, 2) and (5, 6) _____

2. (3, 8) and (1, 4) _____

3. (−5, 1) and (0, 0) _____

B. Write the equation for each line.

1. _____

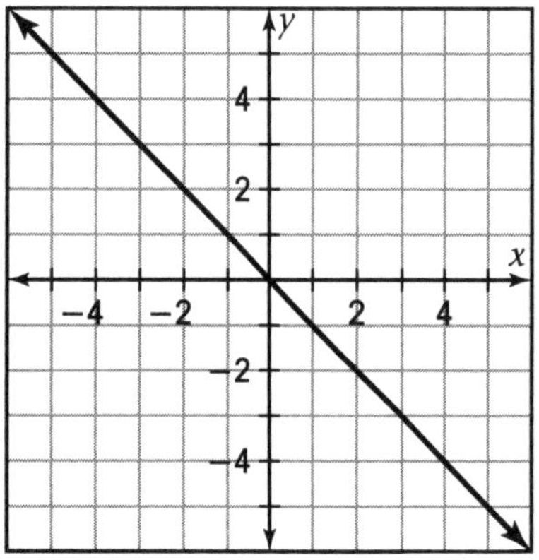

2. _____

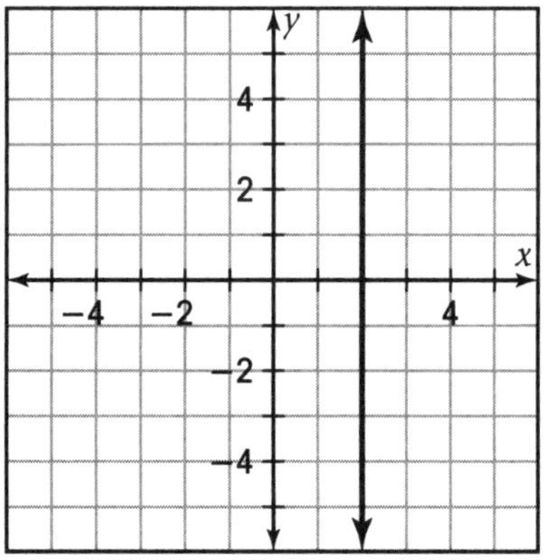

3. _____

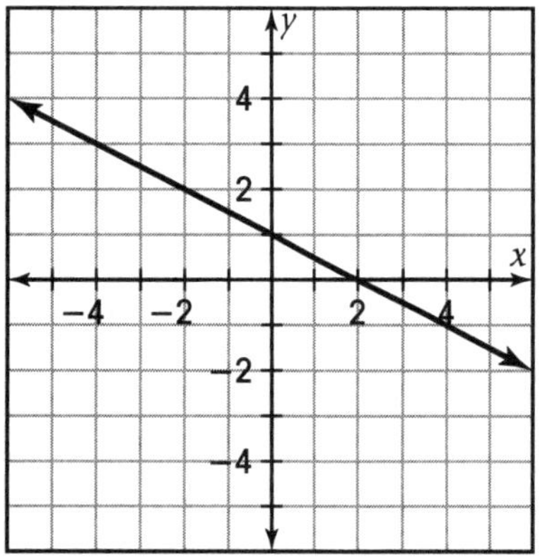

4. _____

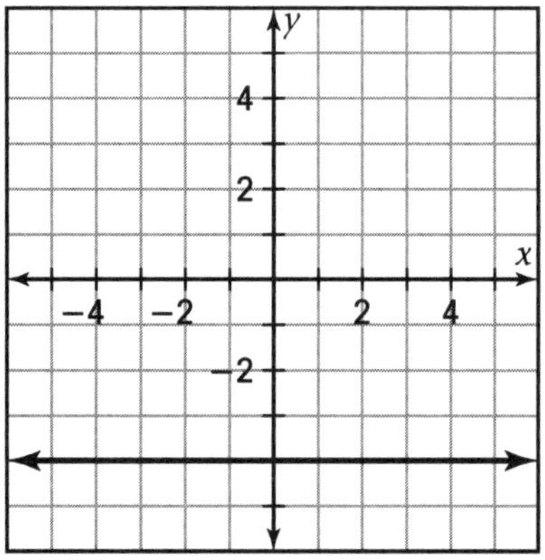

Chapter 6 • Writing Linear Equations   33

Copyright © by Globe Fearon Inc. All rights reserved.

Name _____ Date _____

6 ▷ Parallel and Perpendicular Lines   Exercise 34

Lessons 6.1, 6.4., and 6.5

A. Find the slope from each equation. Then write the slope of a perpendicular line and of a parallel line.

1. $y = 2x + 2$ slope of perpendicular line _____

   slope _____   slope of parallel line   _____

2. $y = -4x$ slope of perpendicular line _____

   slope _____   slope of parallel line   _____

3. $y = x - 3$ slope of perpendicular line _____

   slope _____   slope of parallel line   _____

B. Match each equation with its graph. Write the correct letter on the line.

_____ 1. $y = -x$

a.

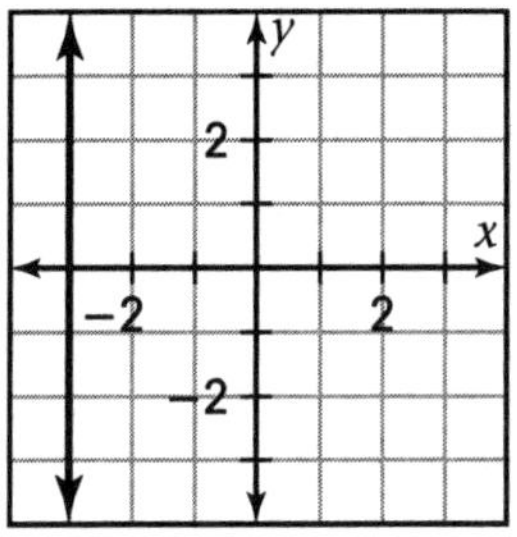

_____ 2. $y = -3$

b.

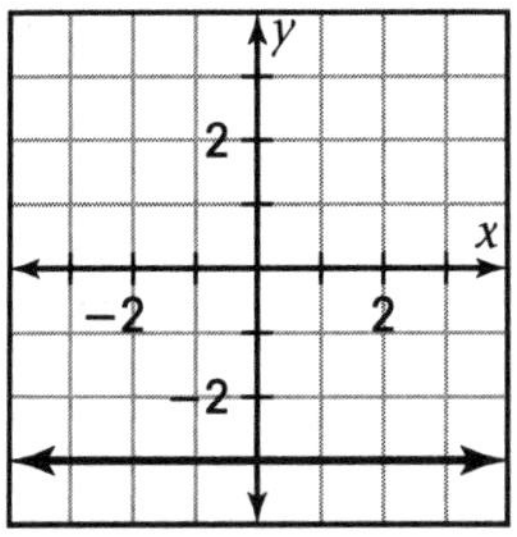

_____ 3. $x = -3$

c.

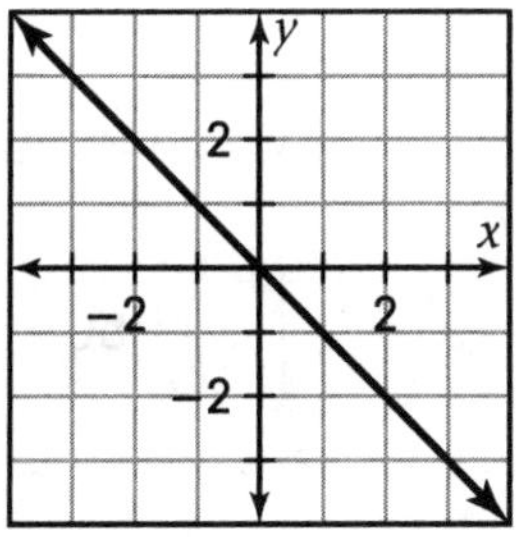

34   Chapter 6 • Writing Linear Equations

Name _____ Date _____

6 ▶ Finding Number Patterns

Exercise 35

Lesson 6.7

A. Describe the pattern in each table. Write your answers on
the lines.

1.

x	y
10	6
11	7
12	8

_____

_____

_____

2.

Pints	Cups
1	2
2	4
3	6

_____

_____

_____

3.

Postage	Total Cost
5	11
6	12
7	13

_____

_____

4.

Students	Teachers
30	1
60	2
90	3

_____

_____

B. Use the table to find a pattern. Then, write the equation
for the pattern.

1.

x	y
8	40
9	45
10	50

equation _____

2.

Height	Coat Length
60	50
61	51
62	52

equation _____

Chapter 6 • Writing Linear Equations   35

Name _____ Date _____

6 ▷ Writing Formulas

Lessons 6.7 and 6.8

Exercise 36

A. Use the table to answer each question.

Minutes	Distance (km)
25	5
50	10
75	15

1. Write an equation that shows how the distance changes with time. Use m for minutes and d for distance.

2. Use the equation to find the distance when $m = 35$.

3. Use the equation to find the distance when $m = 0$.

B. The table below shows how profit at a toy company changes with sales.

Sales	Profit
$12	$4
$18	$6
$24	$8

1. Write an equation to show how profit changes with sales. Use s for sales and p for profit.

2. Use the equation to find the profit when $s = 42$.

3. Use the equation to find the profit when $s = 60$.

36   Chapter 6 • Writing Linear Equations

Name _____ Date _____

7 ▶ Graphing Inequalities

Exercise 37

Lesson 7.1

A. What is wrong with the graph of each inequality? Write your answer on the lines.

1. $x > 6$

[number line: 1 2 3 4 5 6 7 8, open circle at 6, shaded to the left]

_____

2. $x \le -3$

[number line: −8 −7 −6 −5 −4 −3 −2 −1, open circle at −3, shaded to the left]

_____

B. Write the inequality described by the graph on the number line.

1. [number line: 0 1 2 3 4 5 6 7, open circle at 5, shaded to the left]

_____

2. [number line: 0 1 2 3 4 5 6 7, open circle at 2, shaded to the right]

_____

3. [number line: −7 −6 −5 −4 −3 −2 −1 0, closed circle at −2, shaded to the left]

_____

4. [number line: −5 −4 −3 −2 −1 0 1 2, open circle at −3, shaded to the right]

_____

5. [number line: −3 −2 −1 0 1 2 3 4, closed circle at 0, shaded to the right]

_____

Copyright © by Globe Fearon Inc. All rights reserved.

Chapter 7 • Inequalities   37

Name _____ Date _____

7 ▶ Solving Inequalities   Exercise 38

Lessons 7.2 to 7.4

A. Name a number in the solution by looking at the graph.
Then check the number.

1. $a + 1 > -2$

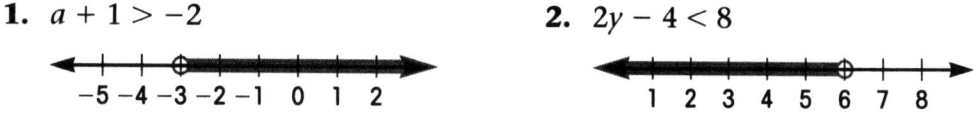

−5 −4 −3 −2 −1 0 1 2

Number in solution _____

Check:

2. $2y - 4 < 8$

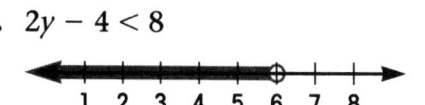

1 2 3 4 5 6 7 8

Number in solution _____

Check:

B. Solve each inequality. Then, graph the solution. Check
with a point from the graph of the solution.

1. $-5m > 10$

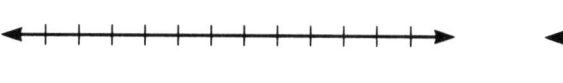

Check:

2. $3w - 3 \leq 6$

Check:

38   Chapter 7 • Inequalities

Copyright © by Globe Fearon Inc. All rights reserved.

Name _____ Date _____

7 ▸ Inequalities with Two Variables   Exercise 39

Lessons 7.1, 7.3, and 7.5

A. Write *true* or *false* after each sentence. If the sentence
is *false,* change the underlined word or words to make
it true.

1. In $x > 5$, 5 <u>is</u> part of the solution.

_____

2. In the graph of an inequality, a <u>solid dot</u> on 3 means the number 3
is a solution of the inequality.

_____

3. Multiplying or dividing both sides of an inequality by the same
negative number <u>does not</u> change the inequality.

_____

4. The "<u>less than</u>" symbol is $<$.

_____

B. Tell whether the point is a solution of the inequality. Write
yes or *no.* Show your work.

1. $(4, 5)$; $y < x + 2$ _____   **2.** $(-2, 1)$; $y \leq x - 1$ _____

3. $(0, 3)$; $3y \geq x$ _____   **4.** $(0, -1)$; $y > 4x - 1$ _____

5. $(6, 2)$; $y > x + 5$ _____   **6.** $(-3, 2)$; $y > 4x$ _____

Chapter 7 • Inequalities   39

Name _____ Date _____

7 ▶ Graphing Solutions in the Coordinate Plane Exercise 40

Lessons 7.5 and 7.6

A. Use the graph to tell whether each point is a solution of the inequality. Write *yes* or *no* on the line.

1. $(4, 5)$ _____

2. $(0, 0)$ _____

3. $(-1, -3)$ _____

4. $(-3, 2)$ _____

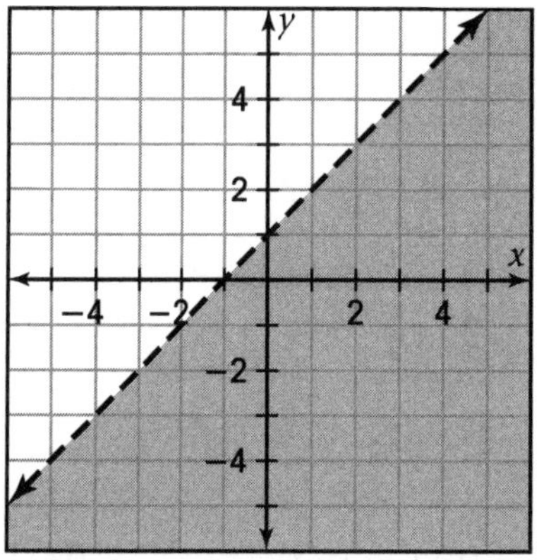

B. Graph the solution of each inequality on a coordinate plane.

1. $y < x$

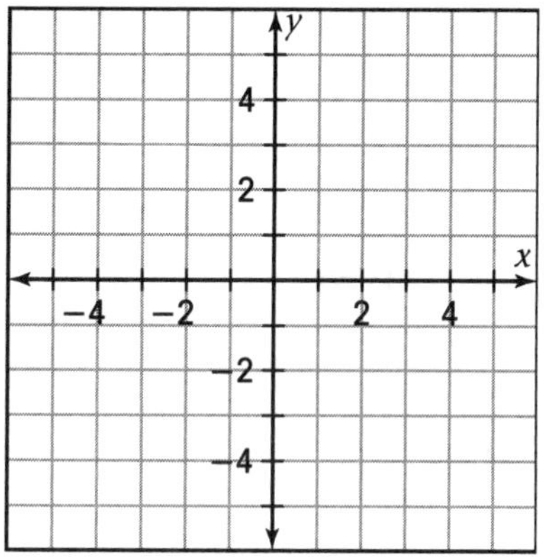

2. $y \geq x - 5$

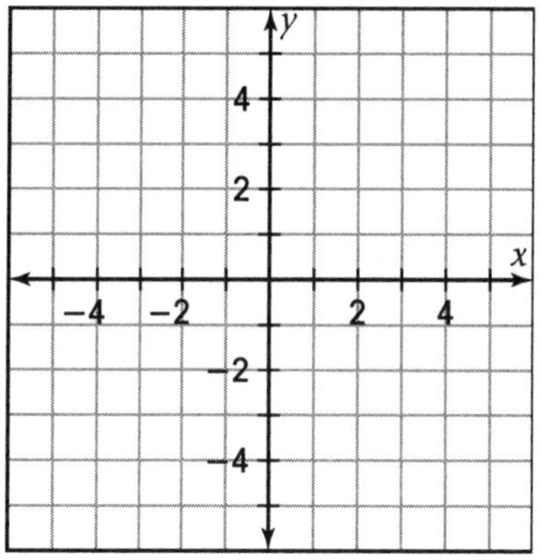

3. $y < \dfrac{-3}{2}x + 5$

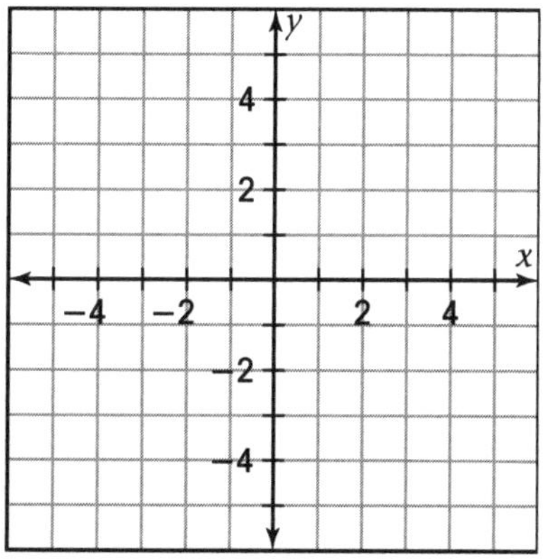

4. $y \leq \dfrac{1}{2}x - 4$

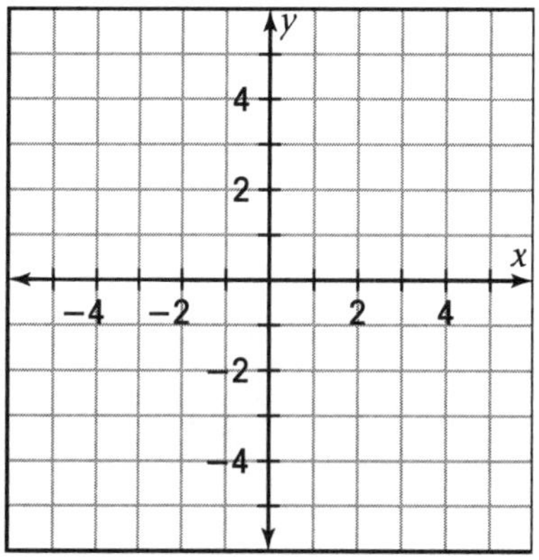

40   Chapter 7 • Inequalities

Copyright © by Globe Fearon Inc. All rights reserved.

Name _____ Date _____

7 ▶ Using Inequalities

Lesson 7.8

Exercise 41

Decide whether each inequality is correct. Write *yes* if it is right. If it is wrong, write the correct inequality.

1. The cookies must bake for no more than 10 minutes. Use *t* for time.

   $t > 10$ _____

2. Tony wants to save at least $16.00. Use *m* for money.

   $m \geq 16$ _____

3. The coldest temperature of the year was less than $-5°C$. Use *t* for temperature.

   $t \leq -5$ _____

CRITICAL THINKING

Write an inequality for each problem. Then, solve the inequality.

1. Two high school students want to buy a tennis racket that costs more than $100.00. If they share the cost evenly, how much will each student pay?

2. Team A has to score at least 55 points to avoid elimination from the game. They now have 43 points. How many more points must Team A score to stay in the game?

3. Jamie is buying tickets to a movie. How many $6.00 tickets can he buy if he wants to spend less than $36.00?

Name _____ Date _____

7 ▶ Using Inequalities

Lesson 7.9

Exercise 42

A. Cathy has $25.00 to spend on clothes for school. She found a clearance sale where each shirt costs $3.00 and each pair of pants costs $5.00. The ordered pair (shirts, pants) tells the number of each item she buys. Use the graph of the inequality to answer the questions $(3s + 5p \leq 25)$.

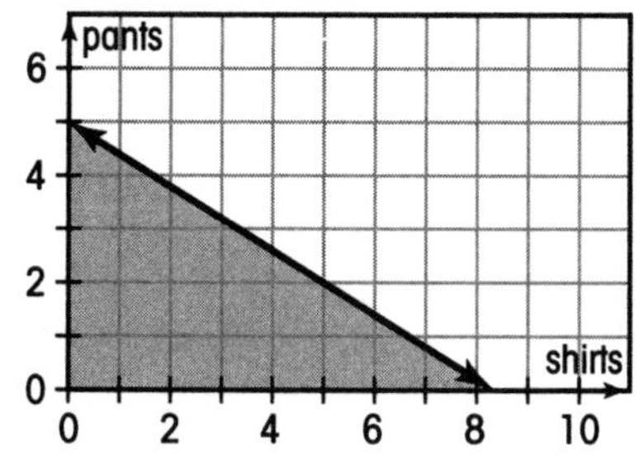

1. Can Cathy buy 6 shirts and 2 pairs of pants?

2. Can she buy 1 shirt and 4 pairs of pants?

3. Can she buy 5 shirts and 2 pairs of pants?

B. Julio has less than $5.00 in coins. All of the coins are quarters and dimes. The ordered pair (quarters, dimes) tells the number of each coin he has. Use the graph of the inequality to answer the questions $(.25q + .10d < 5)$. Explain your answer.

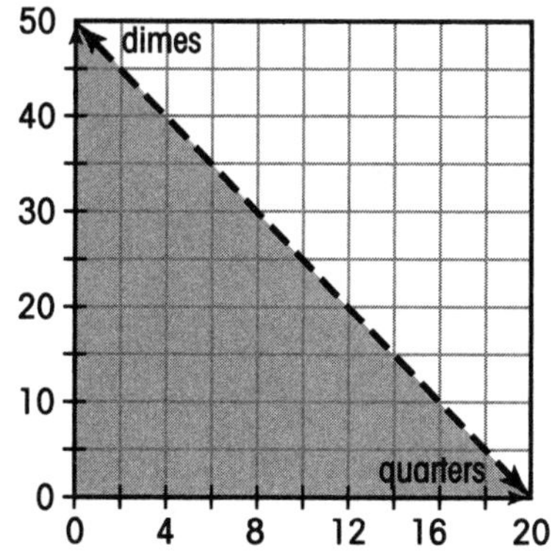

1. Could Julio have 10 quarters and 20 dimes? Why?

2. Could he have 14 quarters and 18 dimes? Why?

3. Could he have 16 quarters and 10 dimes? Why?

42   Chapter 7 • Inequalities

Name _____ Date _____

8 ▸ Systems of Equations and Inequalities   Exercise 43

Lessons 8.1, 8.2, and 8.7

A. Tell whether the ordered pair is the solution of the system
of equations. Write *yes* or *no*.

1. $(5, 1);\ y = x + 7$

$\quad\quad\ y + 2x - 6$

2. $(-1, -4);\ y = x - 3$

$\quad\quad\quad\quad\ y = -5 - x$

3. $(0, 6);\ y = x + 6$

$\quad\quad\ y = 4x + 1$

4. $(3, -5);\ y = 4 - 3x$

$\quad\quad\quad\quad\ y = x - 8$

B. Tell whether the ordered pair is a solution of the system
of inequalities. Write *yes* or *no*.

1. $(-1, 3);\ y > 2x$

$\quad\quad\ y < x - 6$

2. $(4, -1);\ y < 3x + 4$

$\quad\quad\quad\quad\ y \geq 7 - 2x$

C. Find the solution of the system by looking at its graph.
Write the ordered pair in the blank.

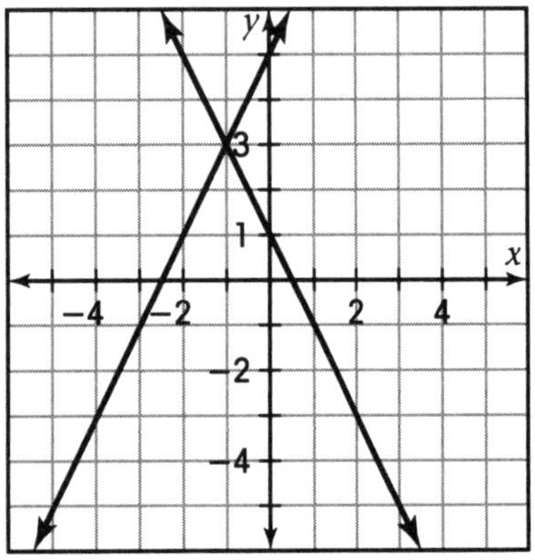

Solution _____

Chapter 8 • Systems of Linear Equations and Inequalities   43

Name _____ Date _____

8 ▸ Solving Systems of Equations   Exercise 44

Lessons 8.3 and 8.4

A. Use the substitution method to solve each system of equations. Then check.

    1. $y = 4x$ **2.** $y = x + 3$

       $y = x + 6$ $x = 4y$

       Check: Check:

B. Use addition or subtraction to solve each system of equations. Then, check.

    1. $6x + y = 10$ **2.** $3x + 5y = -20$

       $2x - y = 6$ $3x + 2y = -17$

       Check: Check:

CRITICAL THINKING

Tell whether you would use substitution or addition to solve each system. Explain why.

    1. $y = 2x + 10$ _____

       $x = 3y$ _____

                       _____

    2. $3x + y = -3$ _____

       $2x - 2y = -10$ _____

                       _____

44   Chapter 8 • Systems of Linear Equations and Inequalities

Name _____ Date _____

8 ▸ Solving Systems by Graphing   Exercise 45

Lessons 8.2 and 8.8

A. Solve by graphing. Write the solution or write *no solution* on the line.

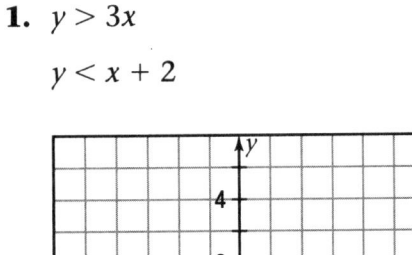

1. $y = 2x$

   $y = x + 5$ _____

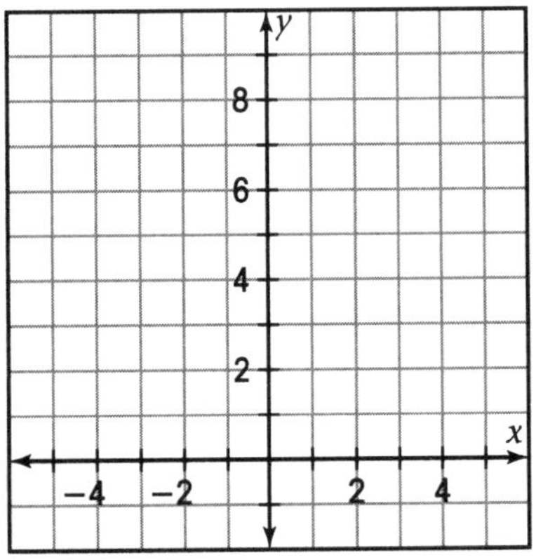

2. $y = 3x$

   $y = x + 2$ _____

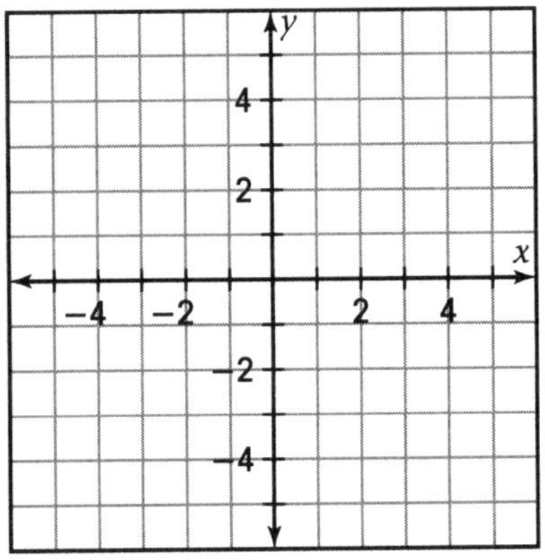

B. Solve by graphing.

1. $y > 3x$

   $y < x + 2$

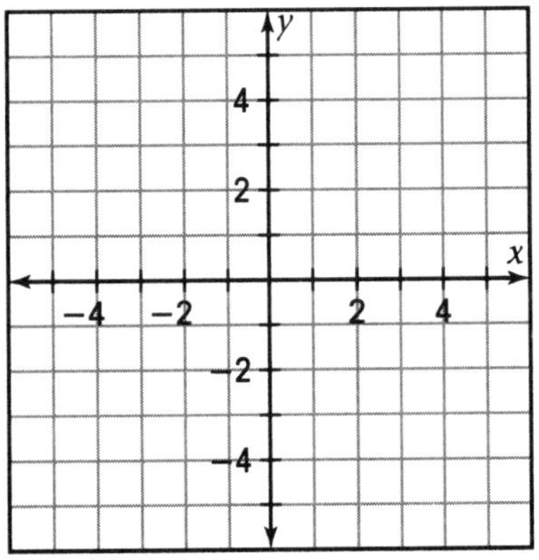

2. $y \geq 2x + 2$

   $y > x - 5$

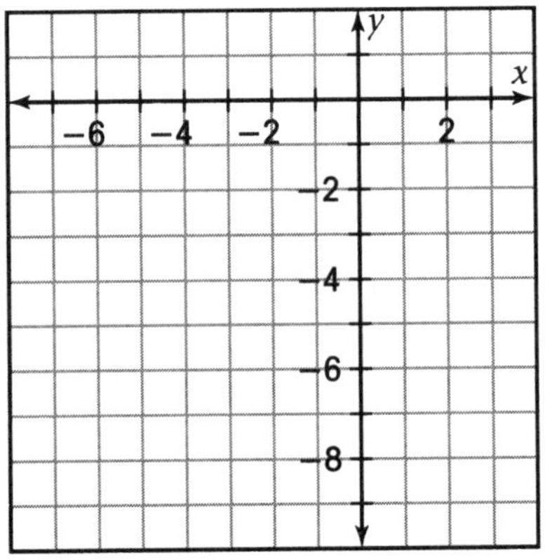

Chapter 8 • Systems of Linear Equations and Inequalities   45

Name _____ Date _____

8 ► Using Systems

Exercise 46

Lesson 8.10

Use a system of equations to solve each problem.

1. Five times a number plus another number is 12. Their difference is 6. What are the two numbers?

   x _____   y _____

2. Last week at the farmer's market, Jerry paid $5.00 for 2 pounds of cucumbers and 1 pound of squash. This week, he paid $9.00 for 3 pounds of cucumbers and 2 pounds of squash. How much does each vegetable cost per pound?

   Cucumbers _____   Squash _____

3. The sum of two numbers is 15. Their difference is 5. What are the two numbers?

   x _____   y _____

4. Maria has 20 coins in her pocket, and all of the coins are either quarters or dimes. The coins total $3.20. How many of each coin does she have?

   Quarters _____   Dimes _____

46   Chapter 8 • Systems of Linear Equations and Inequalities

Name _____ Date _____

8 ▶ Solving Systems

Exercise 47

Lessons 8.5, 8.6, and 8.8

A. Solve each system.

1. $5x + y = -4$

$x + 2y = -17$

2. $8x - 2y = 14$

$-7x + y = 5$

3. $2x + 3y = 15$

$5x + 2y = 21$

4. $4x + 3y = 2$

$3x + 4y = -2$

B. Solve by graphing.

1. $y \geq 5$

$y \leq 3x + 2$

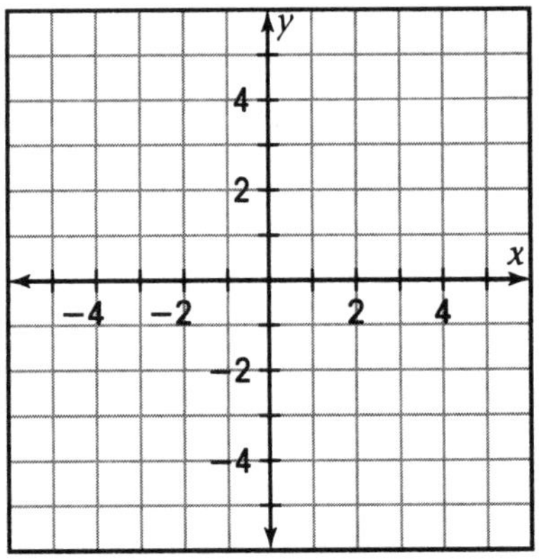

2. $y > x - 3$

$y > 2x - 1$

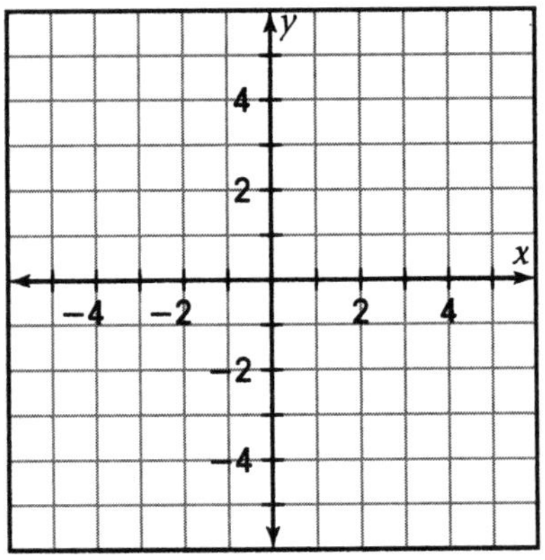

Chapter 8 • Systems of Linear Equations and Inequalities   47

Name _____ Date _____

8 ▶ Using Systems

Exercise 48

Lesson 8.11

A. Write the inequalities.

1. Pamela is planning her Saturday schedule. She wants to sleep at least 8 hours, and she must work at least 5 hours. She can only sleep and work for a total of 16 hours if she plans to run all of her errands.

2. An interior designer orders no more than 100 rolls of red and blue wallpaper. She needs at least 60 rolls of red wallpaper and at least 25 rolls of blue wallpaper.

B. Write the inequalities. Then solve.

A store manager wants to have at least 1,000 records and books in his store. It costs $10.00 to order a record and $6.00 to order a book. He needs at most 600 records and 500 books.

1. Write the inequalities for books, records, and total books and records.

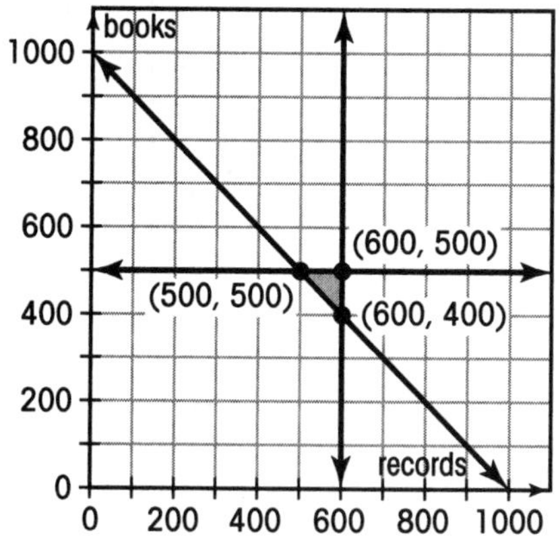

2. Use the graph of the inequalities and the equation Cost = $10r + 6b$ to find the minimum cost.

48   Chapter 8 • Systems of Linear Equations and Inequalities

Copyright © by Globe Fearon Inc. All rights reserved.

Name _____ Date _____

9 ▶ Representing Data   Exercise 49

Lessons 9.1 to 9.3

A. Find the mean, mode, and median of each data set.

1. The classes at Southside High School are not all the same size. The principal wants to give a report to the school board. The numbers of students in each class are as follows: 32, 23, 28, 15, 17, 21, 9, 7, 28.

   Mean _____

   Mode _____

   Median _____

2. The low temperatures during the month of January broke all records. The temperatures (in degrees Fahrenheit) for the first 6 days of the month were as follows: −3, 8, −5, 1, 8, −3.

   Mean _____

   Mode _____

   Median _____

B. Find the minimum, maximum, and range of the set of data.

1. The college basketball team did not have a very consistent season. The team scores for the home games were as follows: 73, 104, 100, 68, 55, 93.

   Minimum _____

   Maximum _____

   Range _____

Chapter 9 • More About Data and Data Analysis   49

Name _____ Date _____

9 ▶ Mean, Mode, Median, Range   Exercise 50

Lessons 9.1 to 9.3

A. Write *true* or *false* after each sentence. If the sentence
is *false,* change the underlined word or words to make
it true.

    1. To find the <u>median</u>, order a set of data from least to greatest.
Then choose the middle number.

    _____

    2. The <u>median</u> is the number that appears most often in a data set.

    _____

    3. The <u>maximum</u> is the largest number in a set of data.

    _____

    4. The <u>range</u> is the difference between the minimum and maximum
values in a set of data.

    _____

    5. For a positive correlation, the data in two sets <u>decrease</u> together.

    _____

    6. The <u>minimum</u> is the smallest number in a set of data.

    _____

B. Tell whether the information about the data is correct.
Write *true* or *false.* If *false,* find the correct value.

Data: 11, 8, 6, 23, 7, 2, 11, 5, 17

    1. mean = 12 _____   **2.** median = 7 _____

    3. mode = 11 _____   **4.** maximum = 17 _____

    5. minimum = 2 _____   **6.** range = 21 _____

50   Chapter 9 • More About Data and Data Analysis

Name _____ Date _____

9 ▶ Displaying Data

Exercise 51

Lessons 9.4 and 9.5

A. Complete the frequency table for the data set of registered voters in a very small town. Then, answer the questions.
R = Republican, D = Democrat, I = Independent
Data set: R, D, R, D, D, I, R, D, I, R, R, R, D, I, R, D, D, I, R, I, R, R, D, I

1.

Voters	Tally	Frequency
Republican		
Democrat		
Independent		
Total		

2. How many voters were registered? _____

3. How many more Republicans are registered than Democrats? _____

4. Which party has the smallest number of registered voters? _____

B. Make a stem-and-leaf plot to display the data set. Then, find the information.

1. 41, 41, 44, 55, 58, 59, 60, 61, 62, 63, 72, 77

Stem	Leaves

2. Find the minimum of the data. _____

3. Find the maximum of the data. _____

4. Find the range of the data. _____

Chapter 9 • More About Data and Data Analysis   51

Name _____ Date _____

9 ▸ Describing Data   Exercise 52

Lesson 9.6

Make a scatter plot for each set of data. Then, tell if the correlation is positive or negative. Write your answer on the line.

1. The table below lists the wind speed and wind chill temperature when the actual temperature is 35° Fahrenheit.

Wind Speed (mph)	Wind Chill (Fahrenheit)
5	33°
10	22°
15	16°
20	12°
25	8°

_____

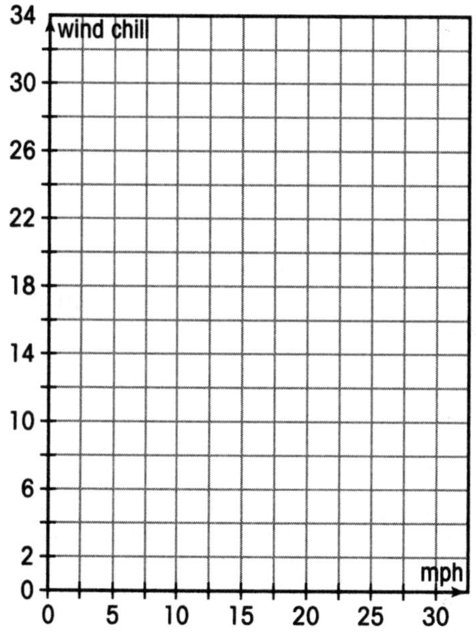

2. The table below lists the height from the sidewalk to the roof and the number of stories from street level of several notable tall buildings.

Height (feet)	Stories
400	32
400	40
529	40
580	50
788	61
859	74
880	63

_____

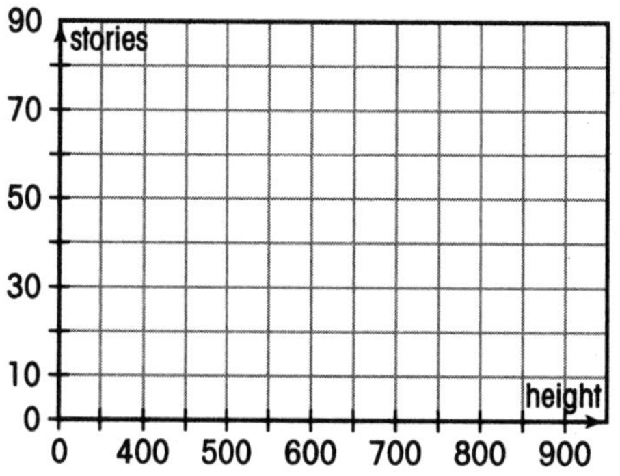

52   Chapter 9 • More About Data and Data Analysis

Name _____ Date _____

9 ▶ Representing Data   Exercise 53

Lesson 9.8

Find the mean, mode, and median. What are the best descriptions for the data sets? Explain why.

1. Jerry's test scores on his last five tests were: 87, 86, 85, 83, 50, 83.

   mean _____ median _____ mode _____

   Best descriptions of the data set _____

   Why? _____

2. The prices of the magazine subscriptions are $21, $6, $10, $17, $3, and $3.

   mean _____ median _____ mode _____

   Best descriptions of the data set _____

   Why? _____

3. The number of cars on local car dealers' lots are 80, 52, 10, 8, and 80.

   mean _____ median _____ mode _____

   Best descriptions of the data set _____

   Why? _____

Chapter 9 • More About Data and Data Analysis   53

Name _____ Date _____

9 ▶ Quartiles   Exercise 54

Lesson 9.9

Find the quartiles of each set of data.

1. Daily stock prices in dollars: $44, $20, $43, $48, $39, $21, $55

   First quartile _____

   Second quartile _____

   Third quartile _____

2. Test scores: 99, 80, 84, 63, 105, 82, 94

   First quartile _____

   Second quartile _____

   Third quartile _____

3. Shoe sizes: 2, 13, 9, 7, 12, 8, 6, 3, 8, 7, 4

   First quartile _____

   Second quartile _____

   Third quartile _____

4. Price of eyeglass frames in dollars: 99, 101, 123, 85, 67, 140, 119

   First quartile _____

   Second quartile _____

   Third quartile _____

5. Number of pets per family: 5, 2, 3, 1, 0, 7, 4, 3, 2, 2, 6

   First quartile _____

   Second quartile _____

   Third quartile _____

54   Chapter 9 • More About Data and Data Analysis

Name _____ Date _____

10 ▶ Powers
Exercise 55

Lesson 10.1

A. Write *true* or *false* after each sentence. If the sentence is *false*, change the underlined word or words to make it true.

1. In the equation $y = 4^x$, 4 is the <u>base</u>.

   _____

2. When the base is positive, the power is always <u>negative</u>.

   _____

3. The product of equal factors is called a <u>power</u>.

   _____

4. In the equation $y = 6^x$, x is the <u>exponent</u>.

   _____

B. Find each power.

1. $(-3)^3$ _____

2. 6^2 _____

3. 5^3 _____

4. $(-1)^5$ _____

C. Find the value of each power when x is -2.

1. x^2 _____

2. x^3 _____

3. x^4 _____

4. x^5 _____

Chapter 10 • Exponents and Functions   55

Name _____ Date _____

10 ▶ Finding Powers   Exercise 56

Lessons 10.2, 10.6 to 10.8

A. Write as factors. Then, rewrite with exponents.

1. $m^4 \bullet m$ Factors _____

   Exponents _____

2. $a^4 b^3 \bullet a^2$ Factors _____

   Exponents _____

3. $rs^3 \bullet 6r^2 s$ Factors _____

   Exponents _____

B. Tell whether each equation is *true* or *false*. If *false*, correct the right side to make a true equation.

1. $r^3 \bullet r^3 = r^9$ _____

2. $x^5 \bullet 3xy = x^2 y^3$ _____

3. $4n^3 p \bullet np^2 = 4n^3 p^2$ _____

C. Multiply or divide.

1. $\dfrac{d^2}{d^2}$ **2.** $x^0 \bullet a^2$ **3.** $a^2 b^0 c \bullet a^0 b^7 c^0$

4. $m^{-3} m^3$ **5.** $n^2 \bullet n^{-1}$ **6.** $a^{-7} b^2 \bullet ab^{-3}$

7. $\dfrac{x^3}{x^5}$ **8.** $\dfrac{a^{-2} b^3}{a^{-3} b}$ **9.** $\dfrac{x^{-5} y^2}{x^2 y^{-7}}$

D. Find each number named in scientific notation.

1. 1.44×10^8 **2.** 6.23×10^{-4} **3.** 2.06×10^{-2}

56   Chapter 10 • Exponents and Functions

Name _____ Date _____

10 ▸ Division and Rules of Exponents

Exercise 57

Lessons 10.3 to 10.5

A. Multiply or divide.

1. $x^5 \bullet x^7$ _____

2. $\dfrac{n^6}{n^3}$ _____

3. $b^4 c^2 \bullet b^5$ _____

4. $\dfrac{a^{10} b^3}{a^5 b^2}$ _____

5. $3x^3 y^2 \bullet 2xy^2$ _____

6. $\dfrac{20n^4 c^6}{4n^4 c^4}$ _____

B. Find the missing term.

1. $c^9 \bullet$ _____ $= c^{16}$

2. _____ $\bullet \, x^2 = x^8$

3. _____ $\bullet \, 2a^3 = 10a^7$

4. $6y \bullet$ _____ $= 18y^2$

5. $a^2 b^3 \bullet$ _____ $= a^4 b^4$

6. _____ $\bullet \, n^2 x^5 = n^3 x^5$

C. Find the missing term.

1. $\dfrac{a^9}{\blacksquare} = a^3$

2. $\dfrac{\blacksquare}{y^2} = y^6$

3. $\dfrac{\blacksquare}{2n^2} = 4n^5$

4. $\dfrac{24x^{10}}{\blacksquare} = 4x^9$

5. $\dfrac{b^5 c^4}{\blacksquare} = b^3 c$

6. $\dfrac{\blacksquare}{y^3} = x^3 y^4$

Chapter 10 • Exponents and Functions   57

Name _____ Date _____

10 ▶ Exponential Functions

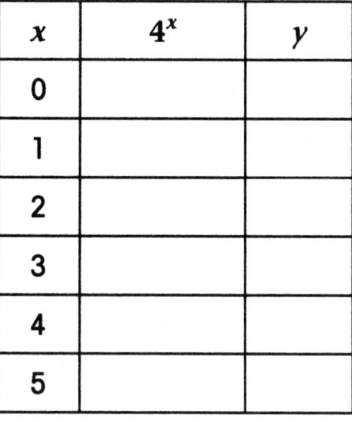

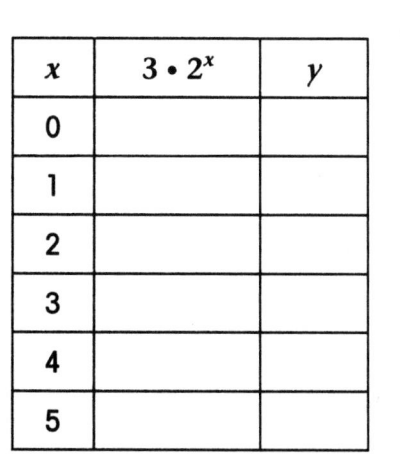

Exercise 58

Lessons 10.9 and 10.10

Find six ordered pairs for each function. Use the values $x = 0$, 1, 2, 3, 4 and 5. Graph each function using the ordered pairs from the table.

1. $y = 3^x$

x	3^x	y
0		
1		
2		
3		
4		
5		

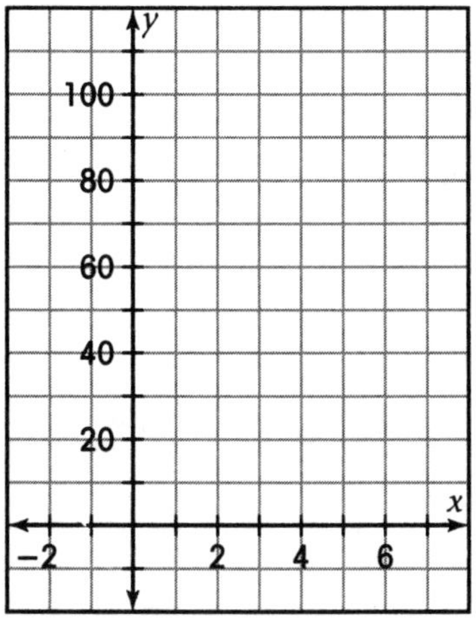

2. $y = 4^x$

x	4^x	y
0		
1		
2		
3		
4		
5		

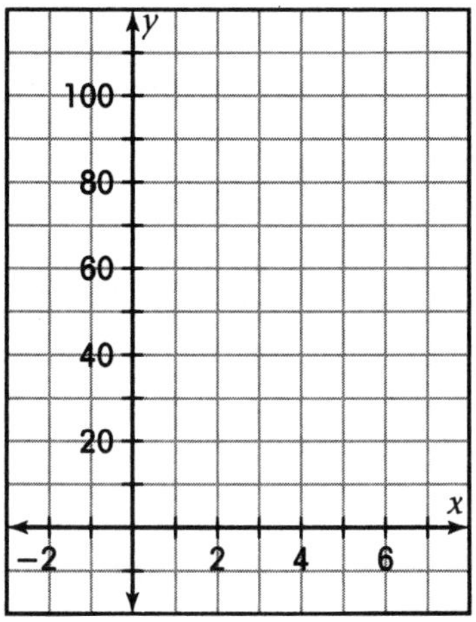

3. $y = 3 \cdot 2^x$

x	$3 \cdot 2^x$	y
0		
1		
2		
3		
4		
5		

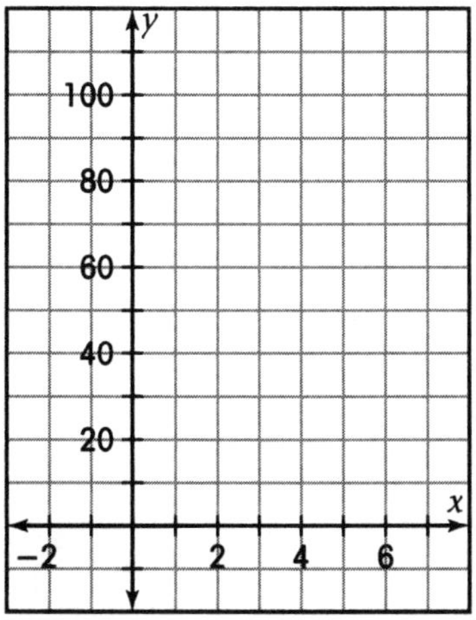

58   Chapter 10 • Exponents and Functions

Name _____ Date _____

10 ▶ Using Tree Diagrams

Exercise 59

Lesson 10.12

Draw a tree diagram to show all the possible choices.

1. There are 3 roads that go from Smithston to Jonesville. There are 2 roads that go from Jonesville to Brown City. How many ways are there to go from Smithston to Brown City if you have to go through Jonesville?

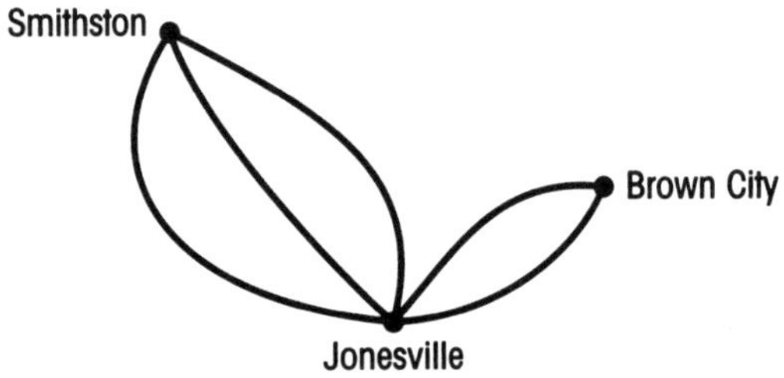

2. The college shop is trying to sell its new line of clothing. It is selling 4 types of pants and 5 styles of tops. How many different outfits can you wear if you buy all of these items?

Chapter 10 • Exponents and Functions   59

Copyright © by Globe Fearon Inc. All rights reserved.

Name _____ Date _____

10 ▸ Using Exponents

Exercise 60

Lessons 10.8 and 10.13

A. The compound interest formula is Total $= a(1 + r)^x$, where
a is the initial deposit, *r* is the rate of interest, and *x* is the
number of time periods the money is invested.

Look at the compound interest equation
$428.72 = 200 \cdot (1.1)^8$ to answer the questions.

1. Find the rate of interest. _____

2. Find the initial deposit. _____

3. Find the number of time periods (years) the money will be in the bank. _____

B. Write the number named by scientific notation.

1. The thinnest piece of glass is 9.8×10^{-4} in. _____

2. A drop of water contains 1.7×10^{22} molecules. _____

3. The growth rate of the abnormal cells in the sample is 1.45×10^5 cells per week.

   _____

4. The weight of the particle being studied was less than expected. It weighed only
   7.823×10^{-3} kg.

   _____

60   Chapter 10 • Exponents and Functions

Name _____ Date _____

11 ▷ Graphing Quadratic Functions

Lesson 11.2

Exercise 61

Make a table of values from the quadratic function.
Then, graph.

1. $y = \frac{1}{2}x^2$

x	$\frac{1}{2}x^2$	y
-4		
-2		
0		
2		
4		

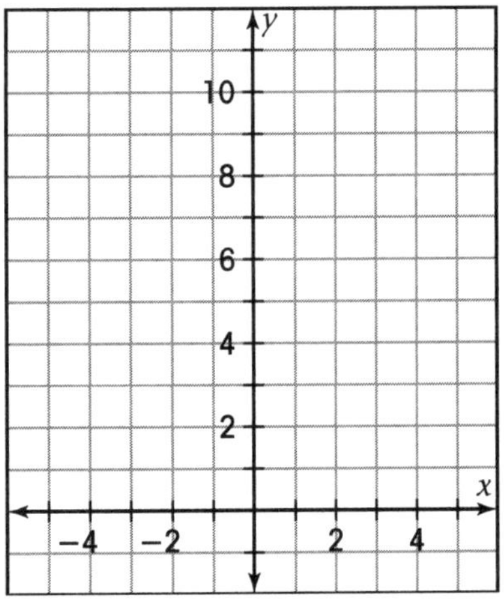

2. $y = x^2 + 1$

x	$x^2 + 1$	y
-3		
-2		
-1		
0		
1		
2		
3		

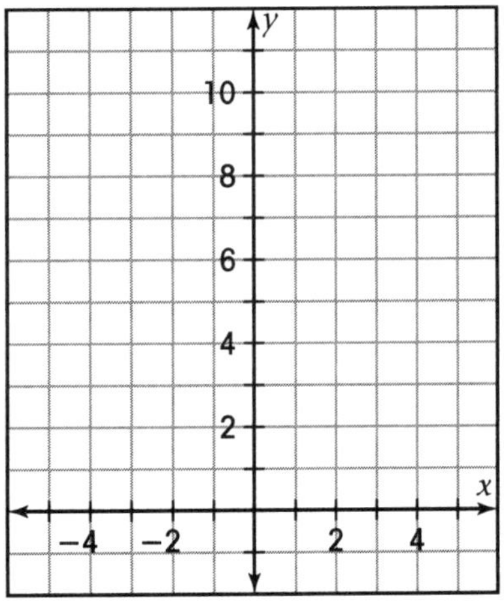

3. $y = -x^2 + 4x$

x	$-x^2 + 4x$	y
-1		
0		
1		
2		
3		
4		
5		

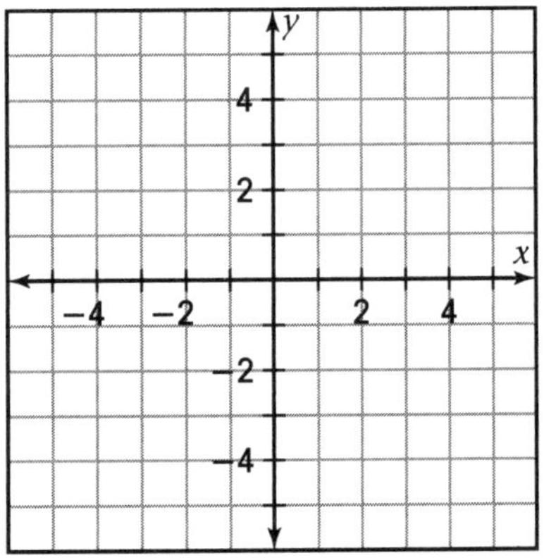

Copyright © by Globe Fearon Inc. All rights reserved.

Chapter 11 • Quadratic Functions and Equations   61

Name _____ Date _____

11 ▶ Minimums and Maximums

Exercise 62

Lesson 11.3

Tell whether the graph has a minimum or maximum.
Then name the point.

1.

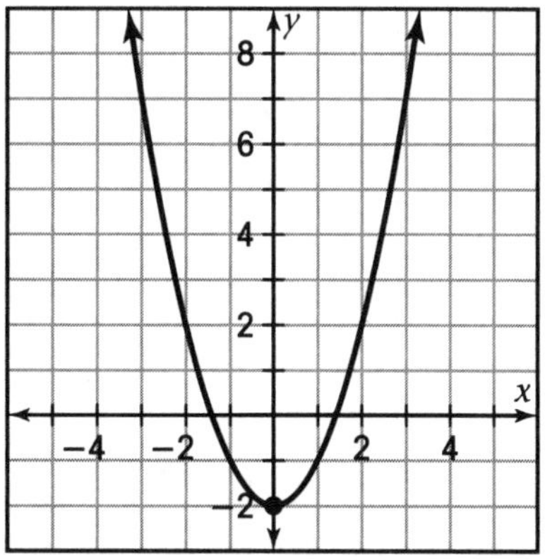

2.

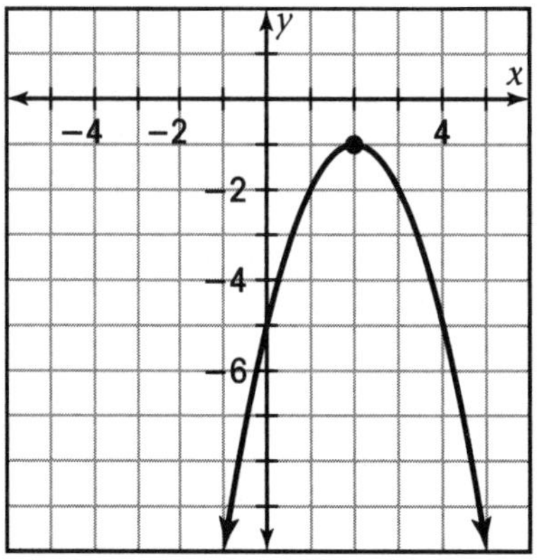

_____ _____

CRITICAL THINKING

The graph below shows part of the graph of a quadratic
function. Complete the graph.

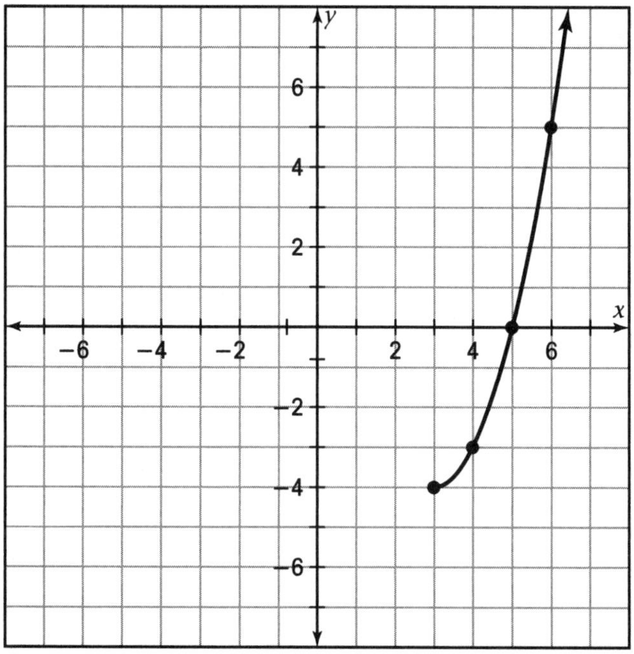

62   Chapter 11 • Quadratic Functions and Equations

Name _____ Date _____

11 ▶ Squares and Square Roots   Exercise 63

Lessons 11.1, 11.3, 11.5, and 11.6

A. Write *true* or *false* after each sentence. If the sentence is *false*, change the underlined word or words to make it true.

1. The equation $y = 3x^2 + 6x + 1$ is a <u>quadratic</u> function.

   _____

2. The <u>maximum</u> of a quadratic function is the largest value of y in a quadratic function.

   _____

3. In a quadratic function, the <u>value of b</u> tells whether the graph opens upward or downward.

   _____

4. <u>Square</u> refers to a number raised to the second power.

   _____

B. Find the square roots of each number.

1. 121 _____   2. 25 _____

3. 144 _____   4. 49 _____

5. 1 _____   6. 100 _____

C. Solve the equations. Check.

1. $x^2 = 64$ 2. $x^2 = 9$

3. $20x^2 = 20$ 4. $x^2 + 7 = 88$

Chapter 11 • Quadratic Functions and Equations   63

Name _____ Date _____

11 ▶ Quadratic Formula

Exercise 64

Lessons 11.4 and 11.7

A. Name the zeros of each function by using its graph.

1.

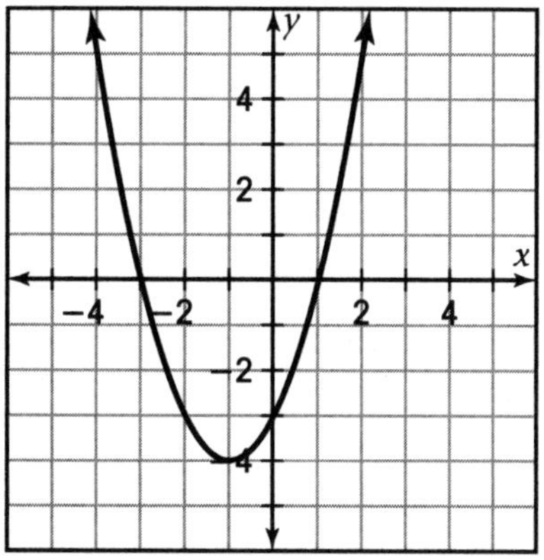

Zeros _____

2.

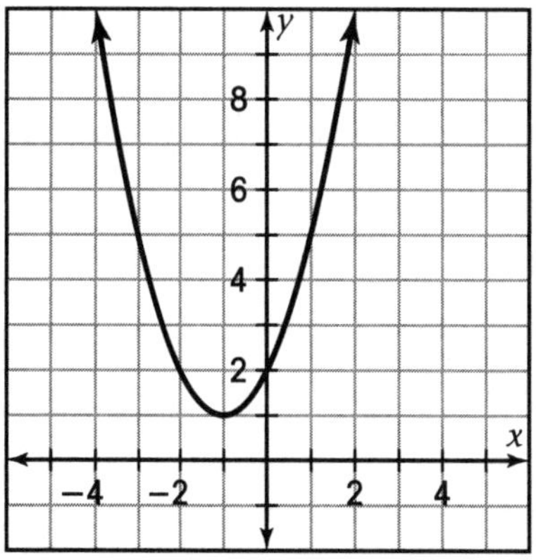

Zeros _____

B. Use the quadratic formula to find the zeros of each equation.

1. $y = x^2 + 2x - 15$

2. $y = x^2 + 4x + 4$

64   Chapter 11 • Quadratic Functions and Equations

Name _____ Date _____

11 ▶ Writing Quadratic Equations

Exercise 65

Lesson 11.9

A. Solve each problem by using a quadratic equation. Use the formula for the area of a square ($A = s^2$).

  1. The area of a square is 64 square inches. Find the length of a side.

  2. A contractor is working on a square patio that will be made of square concrete blocks. He has 25 blocks. How many blocks will there be along each side?

B. Solve each problem by using a quadratic equation. Use the formula for the area of a rectangle ($A = lw$).

  1. The area of a rectangle is 162 square centimeters. The length of the rectangle is 2 times the width. Find the length and width of the rectangle.

  2. An architect is designing a rectangular meeting room so that the length is 3 times the width. She needs the area of the room to be 4,800 square feet. How long should she make each side?

Copyright © by Globe Fearon Inc. All rights reserved.

Chapter 11 • Quadratic Functions and Equations   65

Name _____ Date _____

11 ▶ Vertical Motion Formula   Exercise 66

Lesson 11.10

Solve each problem using the vertical motion formula:
$h = -16t^2 + vt + s$.

1. You drop a penny into a wishing well. After falling for 3 seconds, the penny hits the bottom of the well. How deep is the well?

2. A rock falls off a mountain that is 1,000 feet high. How much time has passed when the rock reaches 744 feet above the ground?

3. A plant falls from a window that is 144 feet high. How much time will it take for the plant to hit the ground?

4. Medical supplies are dropped out of a plane. After 10 seconds, the parachute opens and the supplies are 5,500 feet above the ground. How high was the plane when the supplies were dropped?

66   Chapter 11 • Quadratic Functions and Equations

Name _____   Date _____

12 ▸ Naming Polynomials   Exercise 67

Lessons 12.1 to 12.3

A. Write *true* or *false* after each sentence. If the sentence
is *false*, change the underlined word or words to make
it true.

1. A trinomial is an expression with <u>one term</u>.

_____

2. A binomial is an expression with <u>three</u> terms.

_____

3. A monomial is an expression with <u>two terms</u>.

_____

4. All polynomials have at least one <u>monomial</u>.

_____

B. Add the polynomials.

1. $2x + 5$ and $3x - 2$

2. $3x^2 - x - 5$ and $x^2 - 5x + 7$

3. $3x^2 - 2x - 9$ and $2x + 8$

4. $x^2 + 6x - 2$ and $x^2 + 6$

C. Subtract the polynomials.

1. $4x + 10$ from $8x + 16$

2. $2y^2 - 2$ from $y^2 - 3y + 7$

3. $5y^2 + y + 8$ from $7y^2 - 4y - 4$

4. $2x^2 + 7x - 4$ from $7x^2$

Chapter 12 • Polynomials and Factoring   67

Name _____ Date _____

12 ▸ Polynomials   Exercise 68

Lessons 12.1, 12.2, 12.4, and 12.5

A. The rectangle below is divided into four smaller rectangles. Use the formula, *Area = lw* to find the area of each rectangle.

```
          x           4
      ┌───────────┬───────────┐
      │           │           │
    x │     A     │     B     │ x
      │           │           │
      ├───────────┼───────────┤
      │           │           │
    3 │     C     │     D     │ 3
      │           │           │
      └───────────┴───────────┘
          x           4
```

1. Find the area of A. _____

2. Find the area of B. _____

3. Find the area of C. _____

4. Find the area of D. _____

5. Use your answers to write a polynomial for the total area of the rectangle.

_____

B. Multiply.

1. $3x(x + 2)$ 2. $2a(a - 8)$

3. $(x + 4)(x + 3)$ 4. $(b + 6)(b - 7)$

5. $(x + 4)(x - 7)$ 6. $(b - 4)(b + 4)$

7. $(c - 3)(c + 6)$ 8. $(d - 5)(d - 4)$

68   Chapter 12 • Polynomials and Factoring

Name _____ Date _____

12 ▶ Factors

Exercise 69

Lessons 12.6 and 12.7

A. Decide whether the first expression is a factor of the second expression. Write *yes* or *no*.

1. x^2; $6x$

2. x^2; x^4y^6

3. $3y$; $7y^3$

4. $2a^4$; $8a^4 + 6a^8$

5. $3x^2y$; $3x^3y - 15x^2y^4$

6. $2ab^2$; $2ab + 4a^2b$

B. Factor by finding the greatest common factor.

1. $5c^2d^3 - 15c^3d$

2. $2a^2b - 12a^2b^4$

3. $14t^2 + 7t^5$

4. $20x^2y^3 - 10x^3y^2$

Chapter 12 • Polynomials and Factoring   69

Name _____ Date _____

12 ▶ Factoring

Exercise 70

Lessons 12.8 to 12.10

Factor as the product of two binomials. Then check by multiplying.

1. $n^2 - 121$

2. $p^2 - 49$

3. $64 - x^2$

4. $100 - q^2$

5. $121 - 22x + x^2$

6. $x^2 + 8x - 9$

7. $a^2 + 14a + 13$

8. $c^2 - 20c + 19$

9. $p^2 - 13p + 22$

10. $n^2 - 2n - 8$

Name _____ Date _____

12 ▶ Zero Products

Exercise 71

Lessons 12.11, 12.12, and 12.14

A. Use the Zero Product Property to solve. Then check.

1. $2a(a - 4) = 0$

2. $n(n + 3) = 0$

3. $(3x - 6)(4x - 8) = 0$

4. $(y + 5)(y - 7) = 0$

B. Solve. Use the quadratic formula or factoring.

1. $x^2 - 9x - 10 = 0$

2. $y^2 + 6y + 8 = 0$

3. $a^2 - 36 = 0$

4. $b^2 - 8b - 20 = 0$

Chapter 12 • Polynomials and Factoring   71

Name _____ Date _____

12 ▸ Using Formulas   Exercise 72

Lesson 12.15

Use the given formula to answer the question.

1. Find the area of a triangle with base = 10 m and
    height = 12 m. Use the formula $A = \dfrac{bh}{2}$. b means base,
    and h means height.

2. Find the number of diagonals in a polygon with 8 sides.
    Use the formula $d = \dfrac{s^2 - 3s}{2}$. d means number of
    diagonals and s means number of sides.

3. Find the height in inches of a ball 5 seconds after it is
    thrown upward. Use the formula $h = 40t - 5t^2$. h means
    height, and t means time.

4. Find the value of c if $a = 3$ inches and $b = 4$ inches.
    Use the formula $c^2 = a^2 + b^2$.

Name _____ Date _____

13 ▸ Simplifying Radicals   Exercise 73

Lessons 13.1 and 13.2

A. Write *true* or *false* after each sentence. If the sentence
is *false,* change the underlined word or words to make
it true.

1. $\sqrt{48}$ is between <u>7 and 8</u>. _____

2. $\sqrt{21}$ is between <u>4 and 5</u>. _____

3. $\sqrt{67}$ is between <u>6 and 7</u>. _____

4. $\sqrt{82}$ is between <u>10 and 11</u>. _____

5. $\sqrt{37}$ is between <u>6 and 7</u>. _____

6. $\sqrt{18}$ is between <u>4 and 5</u>. _____

B. Describe the mistake in the work. Then simplify correctly.

1. $\sqrt{45} = \sqrt{9 \cdot 5} = \sqrt{9} \cdot \sqrt{5} = 9\sqrt{5}$ $\sqrt{45}$ _____

   Mistake:

2. $\sqrt{75} = \sqrt{3 \cdot 25} = \sqrt{3} \cdot \sqrt{25} = 3\sqrt{5}$ $\sqrt{75}$ _____

   Mistake:

3. $\sqrt{48} = \sqrt{4 \cdot 12} = \sqrt{4} \cdot \sqrt{12} = 2\sqrt{12}$ $\sqrt{48}$ _____

   Mistake:

Chapter 13 • Radicals and Geometry   73

Name _____ Date _____

13 ▶ Operations with Radicals

Exercise 74

Lessons 13.3 and 13.4

A. Simplify each sum or difference.

1. $2\sqrt{3} + 4\sqrt{3}$ _____

2. $3\sqrt{5} - \sqrt{5}$ _____

3. $4\sqrt{6} + 3\sqrt{7}$ _____

4. $\sqrt{24} + \sqrt{6}$ _____

5. $2\sqrt{50} - 5\sqrt{2}$ _____

6. $\sqrt{24} - \sqrt{20}$ _____

B. Simplify each product.

1. $7\sqrt{2} \cdot 3\sqrt{3}$ _____

2. $5\sqrt{5} \cdot 8\sqrt{7}$ _____

3. $\sqrt{5} \cdot 3\sqrt{10}$ _____

4. $-\sqrt{3} \cdot \sqrt{6}$ _____

C. Simplify each quotient.

1. $\dfrac{10\sqrt{2}}{6}$ _____

2. $\dfrac{4\sqrt{50}}{2\sqrt{5}}$ _____

3. $\dfrac{6\sqrt{3}}{\sqrt{36}}$ _____

4. $\dfrac{7\sqrt{14}}{\sqrt{2}}$ _____

74   Chapter 13 • Radicals and Geometry

Name _____ Date _____

13 ▶ Solving Equations

Lessons 13.5 and 13.7

Exercise 75

A. Solve. Then check.

1. $\sqrt{a} = 13$

2. $2\sqrt{2g} = 4$

3. $\sqrt{x} - 5 = 11$

4. $\sqrt{p - 10} = 12$

B. Use the Pythagorean theorem to solve for the missing side.

1.

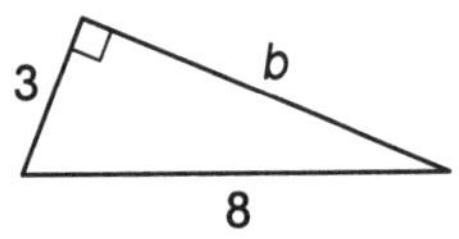

$b =$ _____

2.

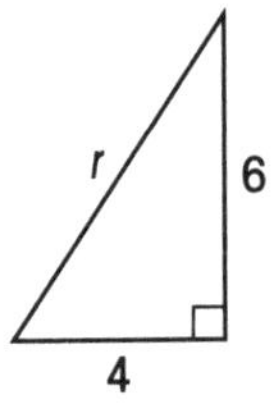

$r =$ _____

3.

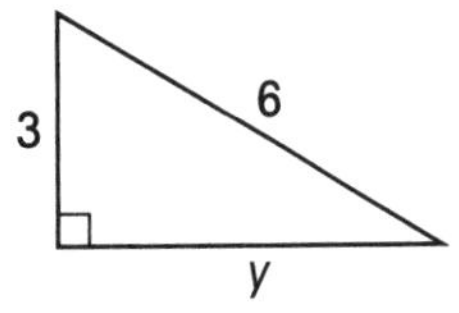

$y =$ _____

4.

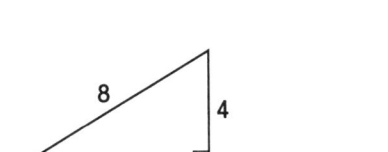

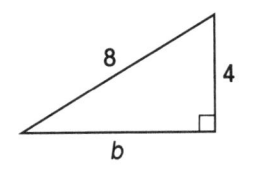

$b =$ _____

Copyright © by Globe Fearon Inc. All rights reserved.

Chapter 13 • Radicals and Geometry   75

Name _____ Date _____

13 ▶ Right Triangles   Exercise 76

Lessons 13.6 to 13.9

A. Circle the correct answer for each question.

1. To use the Pythagorean Theorem, what information must be given?

   a. length of hypotenuse   **b.** length of any two sides

   c. length of short leg   **d.** length of long leg

2. Which equation would you use to find the hypotenuse of a 45°–45°–90° right triangle?

   a. hypotenuse = leg • $\sqrt{2}$ **b.** hypotenuse = (leg)^2

   c. hypotenuse = leg • $\sqrt{3}$ **d.** hypotenuse = 2 • leg

3. Which equation would you use to find the long leg of a 30°–60°–90° right triangle?

   a. long leg = short leg • $\sqrt{2}$ **b.** long leg = (short leg)^2

   c. long leg = short leg • $\sqrt{3}$ **d.** long leg = 2 • short leg

B. Write the equation you would use to find *x*. Then, solve the equation.

1.

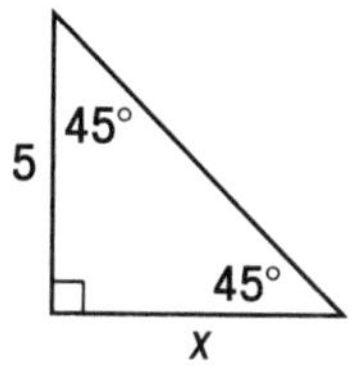

equation _____

solution _____

2.

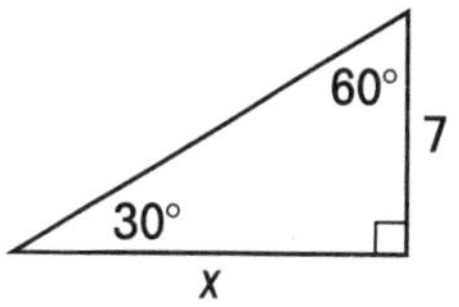

equation _____

solution _____

3.

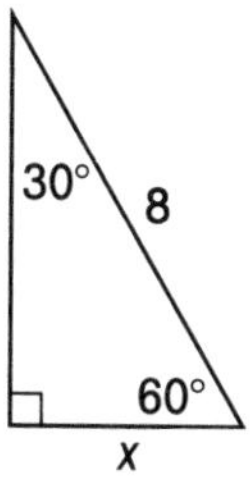

equation _____

solution _____

4.

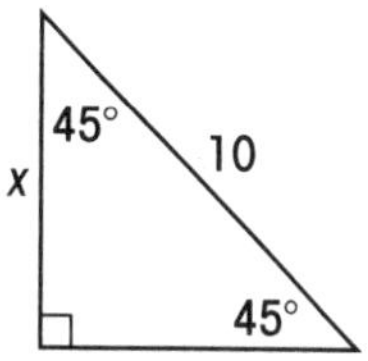

equation _____

solution _____

76   Chapter 13 • Radicals and Geometry

Name _____ Date _____

13 ▶ **Right Triangles**   **Exercise 77**

Lessons 13.6 to 13.9

A. Find the missing sides of each triangle.

1.

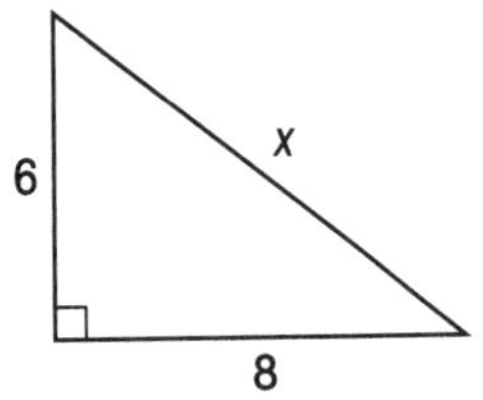

6

x

8

2.

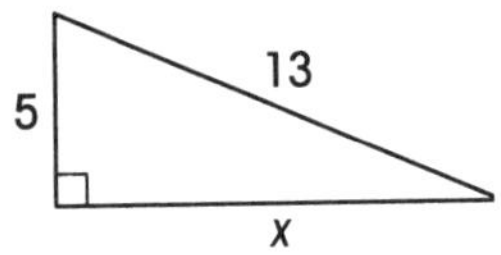

5

13

x

3.

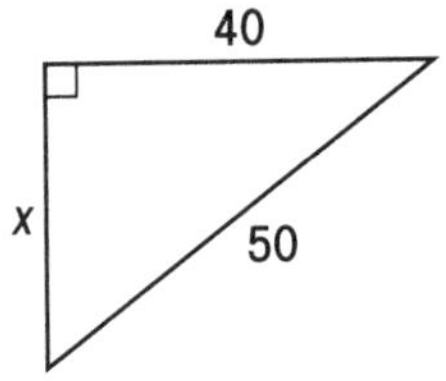

40

x

50

B. Describe what is wrong with each picture.

1.

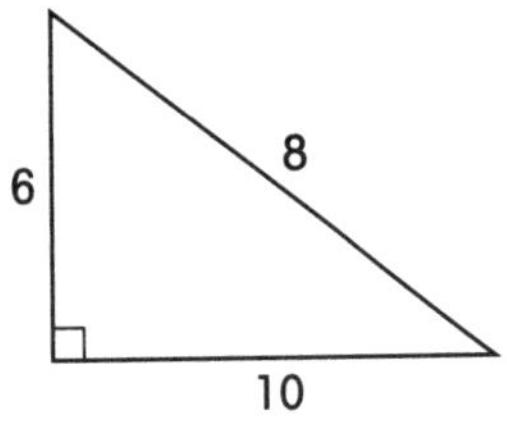

6

8

10

_____

_____

_____

_____

2.

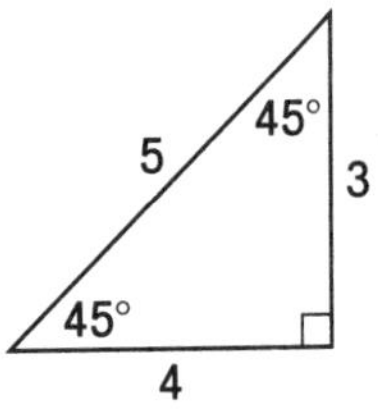

5

45°

3

45°

4

_____

_____

_____

_____

3.

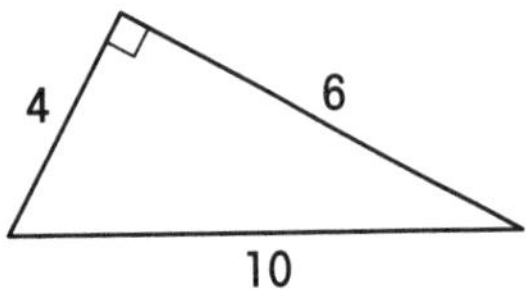

4

6

10

_____

_____

_____

_____

Chapter 13 • Radicals and Geometry   77

Name _____ Date _____

13 ▶ Using the Pythagorean Theorem

Lessons 13.7 and 13.11

Exercise 78

A. Tell whether each triangle is a right triangle. Use the sides and the Pythagorean Theorem. Write your answer on the line.

1.

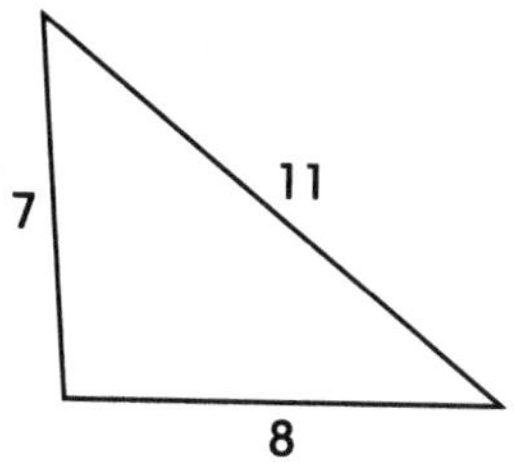

7, 11, 8

_____

2.

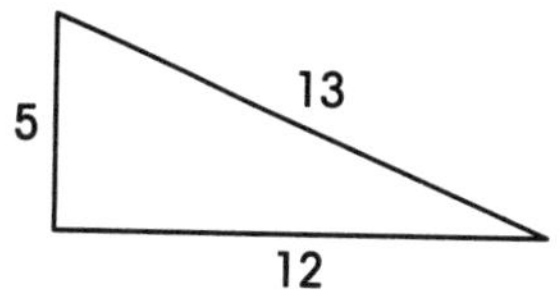

5, 13, 12

_____

B. Use the Pythagorean Theorem to tell whether each triangle is a right triangle. Then write *yes* or *no*. Show your work.

1. Is the triangle formed by the ladder, the wall, and the ground a right triangle?

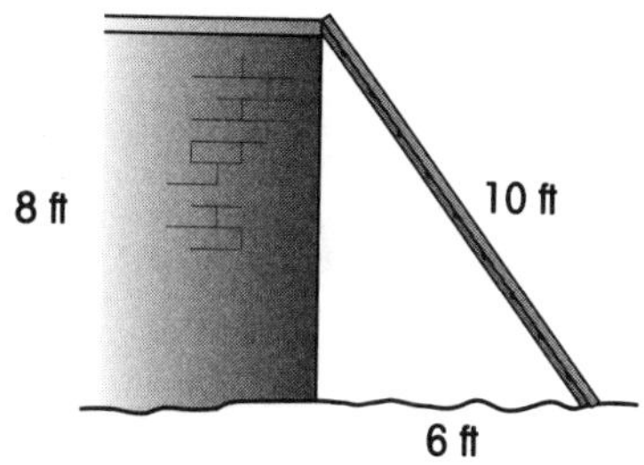

8 ft   10 ft

6 ft

2. Is the triangle formed by the tree and its shadow a right triangle?

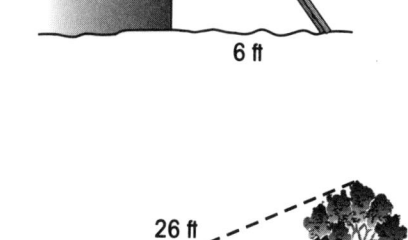

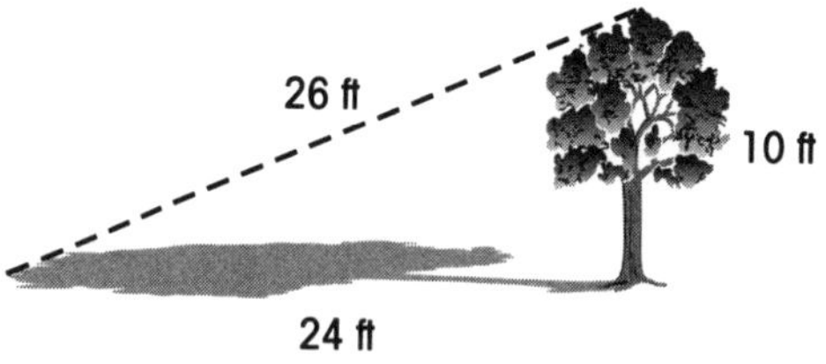

26 ft   10 ft

24 ft

78   Chapter 13 • Radicals and Geometry

Copyright © by Globe Fearon Inc. All rights reserved.

Name _____ Date _____

13 ▶ Finding Distances   **Exercise 79**

Lesson 13.12

A. Graph each pair of points. Then find the distance between the points.

1. $(-3, 4)$ and $(3, 2)$

$d =$ _____

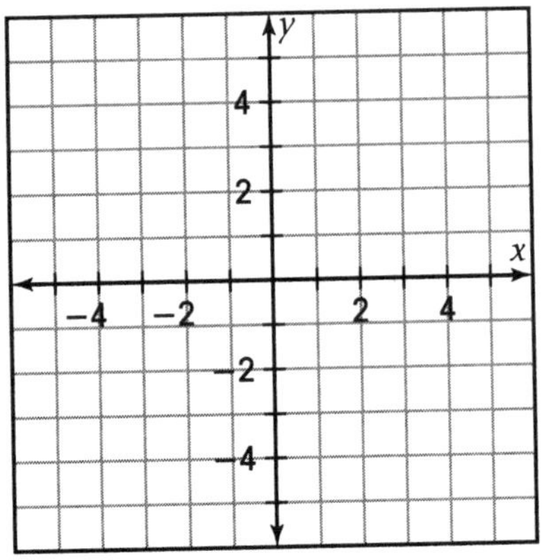

2. $(1, 0)$ and $(5, 3)$

$d =$ _____

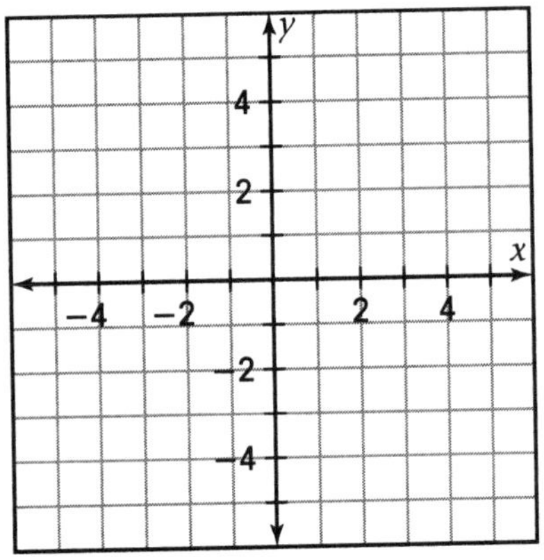

B. Use the picture at the right to find the following distances.

1. Find the distance between points A and B.

$d =$ _____

2. Find the distance between points B and C.

$d =$ _____

3. Find the distance between points C and D.

$d =$ _____

4. Find the distance between points D and A.

$d =$ _____

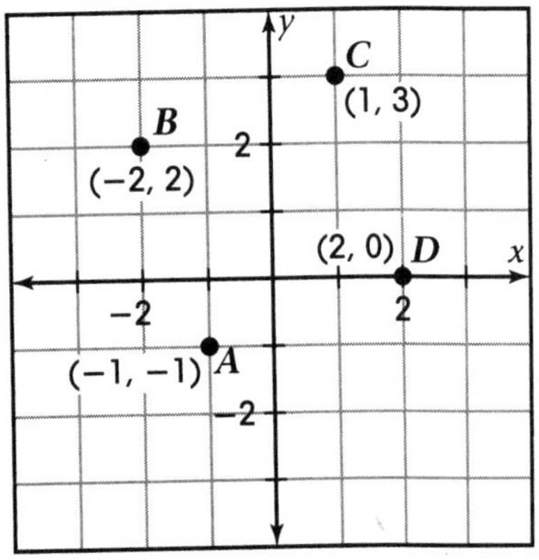

Copyright © by Globe Fearon Inc. All rights reserved.

Chapter 13 • Radicals and Geometry   79

Name _____ Date _____

14 ▶ Rational Numbers and Expressions   Exercise 80

Lessons 14.1 and 14.2

A. Evaluate each expression.

1. $\dfrac{4+y}{2}$ when y is 10.

2. $\dfrac{6}{x+3}$ when x is 9.

3. $\dfrac{4x}{x^2+1}$ when x is 10.

4. $\dfrac{5p}{p^2-3}$ when $p = 1$.

5. $\dfrac{x+7}{6y}$ when x is 2 and y is 3.

6. $\dfrac{6}{a+b}$ when a is 3 and b is 7.

B. State the values of the variable for which the expression is undefined.

1. $\dfrac{b+4}{b-8}$ _____

2. $\dfrac{2a}{(a+2)(a-2)}$ _____

3. $\dfrac{4}{y^2-5y}$ _____

4. $\dfrac{7}{2x^2+2x}$ _____

80   Chapter 14 • Rational Expressions and Equations

Name _____ Date _____

14 ▸ Simplifying Rational Expressions   Exercise 81

Lessons 14.3 and 14.4

A. Simplify all the expressions. Then, draw a line to connect each expression on the left with its equivalent expression on the right.

1. $\dfrac{20}{24}$

        a. $\dfrac{a}{4a^2}$

2. $\dfrac{3a}{12a^2}$

        b. $\dfrac{a^2 - 5a}{4a}$

3. $\dfrac{15}{5(b - 4)}$

        c. $\dfrac{5a^4 b}{6a^4 b}$

4. $\dfrac{4a - 20}{16}$

        d. $\dfrac{18}{6}$

5. $\dfrac{3b + 12}{b + 4}$

        e. $\dfrac{6a + 18}{6}$

6. $\dfrac{a^2 + 7a + 12}{a + 4}$

        f. $\dfrac{3b + 12}{b^2 - 16}$

B. Find the least common multiple.

1. 8 and 24 _____

2. $6n^2$ and $10m^3$ _____

3. $3xy$ and $12x^2$ _____

4. $x - 2$ and $x + 3$ _____

5. $a + 3$ and $a + 6$ _____

6. $r - 5$ and $r + 5$ _____

Chapter 14 • Rational Expressions and Equations   81

Name _____ Date _____

14 ▶ Rational Expressions

Lessons 14.5 to 14.8

Exercise 82

A. Add or subtract as indicated.

1. $\dfrac{4}{3n} + \dfrac{2}{3n}$

2. $\dfrac{3}{5} + \dfrac{3}{10}$

3. $\dfrac{3}{8} - \dfrac{2}{9}$

4. $\dfrac{3}{c+2} - \dfrac{1}{c+2}$

5. $\dfrac{5}{st} + \dfrac{4}{s}$

6. $\dfrac{5b}{12} + \dfrac{b}{8}$

7. $\dfrac{7}{w} - \dfrac{2}{w+4}$

8. $\dfrac{5r}{r+2} + \dfrac{3}{r-2}$

B. Multiply or divide as indicated. Simplify your answer, if possible.

1. $\dfrac{3}{4} \div \dfrac{9}{8}$

2. $\dfrac{3}{4} \div \dfrac{6b}{5a}$

3. $\dfrac{4}{b} \cdot \dfrac{3}{8ab}$

4. $\dfrac{m}{4} \cdot \dfrac{m}{-3}$

5. $\dfrac{r^2}{7s^2} \div \dfrac{3r}{28s}$

6. $\dfrac{uv}{u^2} \cdot \dfrac{uv^2}{v}$

7. $\dfrac{2y+4}{7} \cdot \dfrac{3y}{y+2}$

8. $\dfrac{y^2+4y+4}{y-2} \div \dfrac{y+2}{y-2}$

9. $\dfrac{3y}{6y+18} \cdot \dfrac{y+3}{5}$

10. $\dfrac{x^2-9}{x-3} \div \dfrac{x+3}{x+9}$

82   Chapter 14 • Rational Expressions and Equations

Name _____ Date _____

14 ▶ Proportions

Exercise 83

Lessons 14.1, 14.9, and 14.10

A. Write *true* or *false* after each sentence. If the sentence is *false*, change the underlined word or words to make it true.

1. A proportion is a statement that two ratios <u>are not</u> equal.

   _____

2. The fraction $\dfrac{2y + 6}{4y + 12}$ in <u>lowest terms</u> is $\dfrac{1}{2}$.

   _____

3. 6 is a <u>rational</u> number.

   _____

4. $\dfrac{1}{x + 4}$ is a <u>rational</u> expression.

   _____

5. A rational equation contains <u>irrational</u> expressions.

   _____

B. Solve each proportion.

1. $\dfrac{6}{x} = \dfrac{1}{2}$

2. $\dfrac{5}{r + 4} = \dfrac{3}{r + 2}$

3. $\dfrac{c - 1}{4} = \dfrac{3c + 2}{2}$

4. $\dfrac{m}{m + 4} = \dfrac{2}{m}$

C. Solve each rational equation.

1. $\dfrac{s + 2}{8} + \dfrac{s}{4} = 7$

2. $\dfrac{3}{w} + \dfrac{2}{3} = \dfrac{9}{w}$

3. $\dfrac{4}{x} - \dfrac{5}{2x} = \dfrac{1}{6}$

4. $\dfrac{3}{5y} - \dfrac{7}{10} = \dfrac{4}{2y}$

Chapter 14 • Rational Expressions and Equations   83

Name _____ Date _____

14 ▸ Using Proportions   **Exercise 84**

Lesson 14.12

Use a proportion to solve each problem.

1. If 3 pounds of coffee cost $12.00, how many pounds can you buy for $20.00?

2. Ron can run 15 miles in 2 hours. If he could run for 6 hours, how many miles would he run?

3. A race car driver can travel 15 miles in 3 minutes. How long would it take the driver to travel 180 miles?

4. You can buy 5 pairs of socks for $8.00. How much would you spend if you bought 20 pairs of socks?

5. A printer can produce 48 books in 3 hours. How many books can be printed in 8 hours?

6. James can wash 2 cars in 45 minutes. How long will it take him to wash 6 cars?

7. Yoli has a business meeting 4 out of 5 days each week. How many meetings will she have in 80 days?

84   Chapter 14 • Rational Expressions and Equations

Name _____ Date _____

14 ▸ Inverse Variation

Lesson 14.13

Exercise 85

CRITICAL THINKING

A. In a camera, the opening that lets light in is measured by the f-stop (f). The focal length is the distance from the lens to the point inside the camera to where an image is focused. The f-stop varies inversely with the diameter of the lens. For a focal length of 100 mm, use $f = \frac{100}{D}$ to find the diameter of the lens for each f-stop.

The table gives some f-stops found on cameras. Complete the table to find the lens diameter (in mm) for a focal length of 100 mm. Round your answers to the nearest tenth.

Focal length: 100 mm					
f-stop	1.4	2	2.8	4	8
Diameter					

B. In an electric circuit, a battery can provide electric energy measured in volts. The battery sends an electrical charge, called current, through the circuit. The current is then used by a lamp, radio, or other device. The lamp device provides resistance. The current (I) varies inversely with the resistance (R) of the circuit. With a power of 120 volts, use the equation $I = \frac{120}{R}$ to answer the questions. Resistance is measured in ohms. Current is measured in amperes (amps).

   1. What is the current when the resistance is 20 ohms?

   2. What is the current when the resistance is 30 ohms?

Chapter 14 • Rational Expressions and Equations   85

Name _____ Date _____

15 ▶ Probabilities

Exercise 86

Lessons 15.2 to 15.7

A. Write *true* or *false* after each sentence. If the sentence is *false*, change the underlined word or words to make it true.

1. A <u>compound</u> event is two or more events together.

   _____

2. A <u>complementary</u> event is an event whose outcome affects the outcome of a second event.

   _____

3. <u>A dependent</u> event is an event whose outcome is not affected by the outcome of another event.

   _____

4. Order is not important in a <u>combination</u>.

   _____

5. The value 1.3 <u>could be</u> the probability of an outcome.

   _____

B. Write whether you would use a permutation or a combination to solve each problem. Then solve.

1. Sarah has 7 bracelets. How many different ways can she choose 2 bracelets to

   wear to school? _____

2. In how many different sequences can 4 people stand in a line? _____

   _____

3. A music competition has 6 finalists. They can win first or second place. How

   many ways can the finalists win? _____

86   Chapter 15 • Topics from Probability

Name _____ Date _____

15 ▶ Permutations and Combinations   Exercise 87

Lessons 15.1 to 15.3

A. Write whether you would solve each problem using a permutation or a combination. Then solve.

1. How many ways can you arrange the letters R, O, C, K, E, T?

2. Joe's company has 5 job openings and 9 applicants. How many ways can he choose 5 people to fill the vacant positions?

3. Seven people are competing in a contest. They can win first, second, or third prize. In how many ways can the finalists win?

4. List all the permutations of the letters in the word MATH.

5. List all the combinations of three letters chosen from the word MATH.

B. Find the total number of choices.

1. Esti has 3 skirts and 5 T-shirts. How many choices does she have to pick a T-shirt and skirt?

2. A phone system has been set up to tell people about school cancellations. The principal calls 5 people who each call 5 more people. Those people then each call 5 people. How many people receive a phone call?

Chapter 15 • Topics from Probability   87

Name _____ Date _____

15 ▶ Probabilities   **Exercise 88**

Lessons 15.4 and 15.5

A. Find each probability.

You select one marble from a jar that has 4 black marbles,
6 gray marbles, and 3 white marbles.

1. What is the probability that you will select
    a black marble?

2. What is the probability that you will select
    a gray marble?

3. What is the probability that you will select
    a green marble?

4. What is the probability that you will select
    a white marble?

5. What is the probability that you will select
    a marble that is *not* gray?

CRITICAL THINKING

Write each probability.

You choose a letter at random from the word SPRAIN.

1. What is the probability that you will pick the letter N?

2. What is the probability that you will pick a vowel?

3. What is the probability that you will pick a letter that
    is *not* I?

88   Chapter 15 • Topics from Probability

Name _____ Date _____

15 ▶ Probability

Lesson 15.4

Exercise 89

CRITICAL THINKING

A. The probabilities below describe what could happen when you pick a marble from a certain jar. Use the probabilities to draw the marbles that could be in the jar.

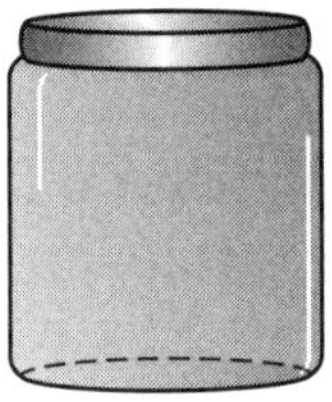

1. P(red marble) = $\frac{3}{9}$ How many red marbles are there? _____

2. P(blue marble) = $\frac{4}{9}$ How many blue marbles are there? _____

3. P(yellow marble) = 0   How many yellow marbles are there? _____

4. P(white marble) = $\frac{2}{9}$ How many white marbles are there? _____

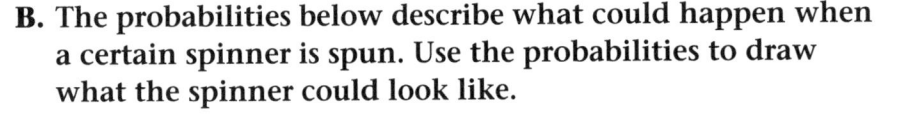

5. How many marbles are in the jar? _____

B. The probabilities below describe what could happen when a certain spinner is spun. Use the probabilities to draw what the spinner could look like.

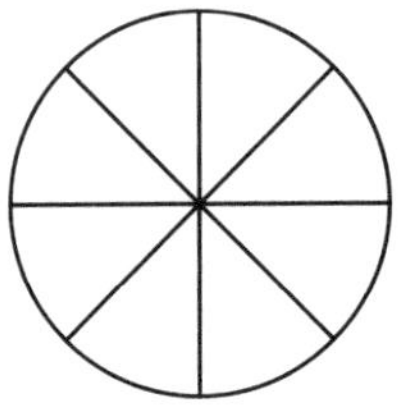

1. P(land on 5) = $\frac{2}{8}$ How many 5s are there? _____

2. P(land on 3) = $\frac{3}{8}$ How many 3s are there? _____

3. P(land on 2) = $\frac{2}{8}$ How many 2s are there? _____

4. P(land on an even number) = $\frac{3}{8}$ How many even numbers are there? _____

5. P(land on a number greater than 5) = $\frac{1}{8}$ How many numbers are greater

    than 5? _____

6. P(land on a number less than 7) = 1   How many numbers are less

    than 7? _____

Chapter 15 • Topics from Probability   89

Name _____ Date _____

15 ▸ Events

Exercise 90

Lessons 15.6 and 15.7

Solve each problem.

1. Allison has a bag containing 6 red chips and 4 black chips. Without looking, she pulls out one chip. Then without putting the first chip back, she pulls out a second chip. What is the probability that both chips are red?

2. You have a spinner with 4 odd numbers and 3 even numbers. What is the probability that the spinner will *not* land on an odd number?

3. You roll a cube with sides numbered 0 to 5. Then, you toss a coin. What is the probability that you roll a 5 and toss heads?

4. Carla is taking a multiple choice test. The probability that she guesses correctly on a question is $\frac{1}{5}$. What is the probability that she will guess incorrectly?

5. You flip a coin 4 times. What is the probability that heads will come up all 4 times?

Name _____ Date _____

15 ▷ Probabilities   **Exercise 91**

Lesson 15.9

Solve each problem. Write the probability as a percent.

1. A survey asks 20 people if they have a pet. Seven people answer "yes." Find the probability that a person has a pet.

2. The team has won 9 of its last 12 games. Find the probability that the team wins.

3. A factory makes computer chips. In a sample of 8,000 chips, 50 of the chips were defective. Find the probability that a chip is defective.

4. A newspaper reports that 21 of the 25 people surveyed voted in the last election. Find the probability that a person voted in the last election.

5. A survey asked 60 people if they preferred apples or oranges. 42 people said they preferred apples. Find the probability that a person prefers apples.

Name _____ Date _____

15 ▶ Predicting Outcomes

Lesson 15.10

Exercise 92

Predict each outcome.

1. The probability of a defective car radio is 7%. How many defective radios would you expect to find in 500 cars?

2. Jim plays basketball. He scores a basket in 75% of his free throws. How many free throws would you expect Jim to make in 40 attempts?

3. There is a 20% chance of rain in the next 12 days. How many days do you expect it to rain?

4. 29% of the people surveyed said they bought a book last week. How many people would you expect to have purchased a book if 1,000 people were surveyed?

5. In one industry, 22% of the products sold become obsolete very quickly. How many products in 450 would you expect to become obsolete very quickly?

6. Maria is a soccer player. She scores 90% of her goal shots. If she takes 8 shots, how many goals can she expect to score?

7. There is a 10% chance of sleet in the next 10 days. How many days do you expect sleet?

92   Chapter 15 • Topics from Probability